EXPERIENCING GEOMETRY

on Plane and Sphere

Geometry

Logic can only go so far —
    after that I must see-perceive-imagine.
This geometry can help.

I may reason logically thru theorem
      and propositions galore,
    but only what I perceive is real.

If after studying I am not changed —
    if after studying I still see the same —
      then all has gone for naught.

Geometry is to open up my mind
    so I may see what has always
          been behind
    the illusions that time
      and space construct.

Space isn't made of point and line
    the points and lines are in the mind.
The physicists see space as curved
    with particles that are quite blurred.
And, when I draw, everything is fat
    there are no points and that is that.
The artists and the dreamer knows
    that space is where an image grows.
For me it's a sea in which I swim
    a formless sea of hope and whim.

Thru my fear of Infinity and One
    I structure space to confine
      my imagination away from the idea
        that all is One.

But, I can from this trap escape —
    I can see the geometry in which I wander
    as but a structure I made to ponder.

I can dare to let go the structures
        and my fears
    and look beyond
    to see what is always there to see.

But, to let go, I must first grab on.
    Geometry is both the grabbing on
    and the letting go.
It is a logical structure
    and a perceived meaning —
    Q.E.D.'s and "Oh! I see!"'s.
It is formal abstractions
    and beautiful contraptions.
It is talking precisely about that
    which we know only fuzzily.
But, in the end, and, most of all,
    it is seeing-perceiving
      the meaning that
        I AM.

— David Henderson, 1978

Stetson University

3 4369 00305683 X

EXPERIENCING GEOMETRY

on Plane and Sphere

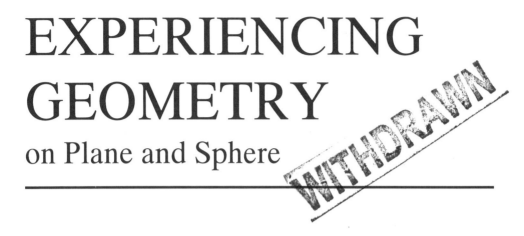

David W. Henderson
Cornell University

with writing input from

Eduarda Moura
and

Justin Collins
Kelly Gaddis
Elizabeth B. Porter
Hal W. Schnee
Avery Solomon

PRENTICE HALL
Upper Saddle River, NJ 07458

DATE DUE			
GAYLORD			PRINTED IN U.S.A.

Library of Congress Cataloging-in-Publication Data

Henderson, David W. (David Wilson), (date)
  Experiencing Geometry : on plane and sphere / by David W. Henderson.
    p.   cm.
  Includes bibliographical references and index.
  ISBN 0-13-373770-5
    1. Geometry.   I. Title.
QA453.H497   1996
516—dc20   95-11072
                                        CIP

Acquisitions editor: George Lobell
Production editor: Barbara Mack
Managing editor: Jeanne Hoeting
Director of production and manufacturing: David W. Riccardi
Cover design: Bruce Kenselaar
Cover photo: Interior of City Hall Civic Center, NYC. Berenholtz Photography
Manufacturing buyer: Alan Fischer

 © 1996 by Prentice-Hall, Inc.
Simon & Schuster/A Viacom Company
Upper Saddle River, NJ 07458

All rights reserved. No part of this book may be reproduced, in any form or by any means, without permission in writing from the publisher.

Portions of this material is based upon work supported by the National Science Foundation under Grant No. USE-9155873 and by Dwight David Eisenhower Title IIA grants administered by the New York State Department of Education. Any opinions, findings, and conclusions or recommendations expressed in this material are those of the author and do not necessarily reflect the views of the National Science Foundation or the New York State Department of Education.

Printed in the United States of America
10 9 8 7 6 5 4 3 2

ISBN 0-13-373770-5

PRENTICE-HALL INTERNATIONAL (UK) LIMITED, LONDON
PRENTICE-HALL OF AUSTRALIA PTY. LIMITED, SYDNEY
PRENTICE-HALL CANADA INC., TORONTO
PRENTICE-HALL HISPANOAMERICANA, S.A., MEXICO
PRENTICE-HALL OF INDIA PRIVATE LIMITED, NEW DELHI
PRENTICE-HALL OF JAPAN, INC., TOKYO
SIMON & SCHUSTER ASIA PTE. LTD., SINGAPORE
EDITORA PRENTICE-HALL DO BRAZIL, LTDA., RIO DE JANEIRO

Dedicated

to all the students who have studied geometry with me
(you have taught me much about geometry)

and to Susan Alida
(you have taught me much about my Self)

Contents

Preface

What geometrician or arithmetician could fail to take pleasure in the symmetries, correspondences and principles of order observed in visible things? Consider, even, the case of pictures: those seeing by the bodily sense the productions of the art of painting do not see the one thing in the one only way; they are deeply stirred by recognizing in the objects depicted to the eyes the presentation of what lies in the idea, and so are called to recollection of the truth — the very experience out of which Love rises.
— Plotinus, *The Enneads*, II.9.16 [A: Plotinus]

The formal expression of "straightness" is a part of differential geometry and is a very difficult formal area of mathematics. However, the concept of "straight," an often used part of ordinary language, is generally used and experienced by humans starting at a very early age. This book will lead the reader on an exploration of the notion of straightness and the closely related notion of parallel on the plane and on the sphere.

This book is based on a junior/senior level course I have been teaching for twenty years at Cornell for mathematics majors, high school teachers, future high school teachers, and others. Most of the chapters start intuitively so that they are accessible to a general reader with no particular mathematics background except imagination and a willingness to struggle with ideas. However, the discussions in the book were written for mathematics majors and mathematics teachers and thus assume of the reader a corresponding level of interest and mathematical sophistication.

The course emphasizes learning geometry using reason, intuitive understanding, and insightful personal experiences of meanings in geometry. To accomplish this the students are given a series of inviting and challenging problems, are encourage to write and speak their reasonings and understandings; and then I listen to and critique their thinking and use it to stimulate the whole class discussions.

Most of the problems are approached both in the context of the plane and in the context of a sphere (and sometimes a cylinder and cone). I find that by exploring the geometry of a sphere my students gain a deeper understanding of the geometry of the plane. For example, the question of whether or not Side-Angle-Side holds on a sphere leads one to pursue the question of what is it about Side-Angle-Side that makes it true on the plane. I also introduce the modern notion of "parallel transport along a geodesic" which is a notion of parallelism that makes sense on both the plane and on a sphere (in fact, on any surface). While exploring parallel transport on a sphere the students are able to more fully appreciate that the similarities and differences between the Euclidean geometry of the plane and the non-Euclidean geometries of a sphere and other spaces are not adequately described by the usual Parallel Postulate. I find that the early interplay between the plane and a sphere enriches all the later topics whether on the plane or on a sphere.

In my course *the distinction between learning activities and assessment activities is blurred*. I present a sequence of problems (together with motivation, discussion of contexts, and connections of the problems with other areas of mathematics and life). I tell the students:

> Write out your thinking to each problem. We will return your papers with comments about your solutions. Respond to our comments — use them as invitations to explore, to clarify your understanding of the problem, or to clarify our understanding of your solution. Keep responding until you understand. Turn in whatever your thinking is on a question even if only to say "I don't understand such and such" or "I'm stuck here"; be as specific as possible. Feel free to ask questions. This will allow the sharing of ideas, and you will benefit more from class sessions.

The students then work on the problems either individually or in small groups and report their thinking back to me and the class. This cycle of *writing, comments, discussion* continues on each problem until both the students and I are satisfied, unless external constraints of time and resources intervene.

What I have discovered is that in this process not only have the students learned from the course, but also I have learned much about geometry from them. At first I was surprised; how could I, the teacher, learn mathematics from the students? But this learning has continued for 20 years and I now expect its occurrence. In fact, as I expect it more and more and learn to listen more effectively to them, I find that a greater portion of my students show something new to me about geometry. I have also discovered that I am learning more (percentage-wise) from those students who differ from me in terms of gender and race. For more discussion of this, see the "Message to the Reader" on pages xx-xxiv.

Useful Supplements

Instructors may obtain from the publisher an *Instructor's Manual* which contains for each problem a full discussion of possible solutions, examples of students' original work, and suggestions for class discussion.

For exploring properties on a sphere it is important that you have a model of a sphere that you can use. Some people find it helpful to purchase plastic sphere sets which include a transparent sphere, a spherical compass, and a spherical "straight edge" which doubles as a protractor. These sets should be available in your bookstore or from Key Curriculum Press, Berkeley, CA. They work well for small group explorations in the classroom. The instructor should also have a large "black board" sphere that can be written on with chalk — these spheres are often common in chemistry classrooms. However, a beach ball or basketball will also work, particularly if used with rubber bands large enough to form great circles on the ball. Students often find it convenient to use worn tennis balls ("worn" because the fuzz can get in the way) because they can be written on and are the right size for ordinary rubber bands to represent great circles.

Acknowledgments

I acknowledge my debt to all the students and teachers who have attended my geometry courses. Most of these people have been students at Cornell or teachers in the surrounding area of upstate New York, but they also include students at Birzeit University in Palestine and teachers in the new South Africa. Without them this book would have been an impossibility.

Starting in 1986, Avery Solomon and I organized and taught a program of inservice courses for high school teachers under the financial support of Title IIA Grants administered by the New York State Department of Education. This is now called the Cornell/Schools Mathematics Resource Program (CSMRP). As a part of CSMRP we started recording classes and writing notes on the material. Some of the material in this book had its origins in those notes, but they never threatened to become a textbook. I thank Avery for his modeling of enthusiastic teaching, his sharp insights, and his insistence on preserving the teaching materials. In addition to Avery, my friends, Marwan Awartani, a professor at Birzeit University, and John Volmink, the director of the Centre for the Advancement of Science and Mathematics Education in Durban, South Africa, have for a long period of time consistently encouraged me to write this book.

A few years ago my colleague Maria Terrell suggested that five of us at Cornell who have been teaching non-traditional geometry courses (Avery Solomon, Bob Connelly, Tom Rishel, Maria, and I) submit a proposal to the National Science Foundation for a grant to write up materials on our courses. The fact that we were awarded the grant (in 1992) is largely due to Maria's persistence, clear thinking, and encouragement. It is this grant which gave me the necessary support to start the writing of this book. I thank the NSF's Program on Course and Curriculum Development for its support.

The major portions of this book were written during the 1992-93 academic year in which I taught the course both semesters. Eduarda Moura was my teaching assistant for these courses. She was supported by the NSF grant to assist me by describing the classroom discussion and the student homework upon which the content of this book is based. Much of this book (and especially the instructor's manual) are derived from her efforts. In addition to Eduarda, Kelly Gaddis, Beth Porter, Hal Schnee, and Justin Collins were also supported by the grant and made significant contributions to the writing of this book. I thank them all for their excellent contributions, their support of my work, and their friendship. The final writing and the decisions as to what to include and what not to include have all been mine, but they have been based on the foundation that was started with Avery and the CSMRP materials and was continued wtih Eduarda, Justin, Kelly, Beth, and Hal during 1992-93.

Since the spring of 1992, the early drafts of the book have been used by me and others at Cornell and 13 other institutions. Various other individuals have worked through the book outside a classroom setting. From these students, instructors and others I have received encouragement and much valuable feedback that has resulted in what I consider to be a better book. In particular, I want to thank the following persons for giving me feedback and ideas which I have used in this final version: David Bray, Douglas Cashing, Helen Doerr, Jay Graening, Christine Kinsey, István Lènárt, Julie Lubell, Richard Pryor, Amanda Cramer and her students, Erica Flapan and her students, Linda Hill and her students, Tim Kurtz and his students, Judy Roitman and her students, Bob Strichartz and his students, and Walter Whitely and his students. Susan Henderson, my wife, spent many hours proof-reading and refining the text and was always my consultant on matters of aesthetics.

The entire production of the manuscript (typing, formatting, drawings, and final layout) has been accomplished using an integrated word processing software under my direction. Finally, I wish to thank George Lobell, Senior Editor at Prentice-Hall, for the vision and enthusiasm with which he shepherded this book through the publication process.

Ithaca, NY, May 1995

David W. Henderson

Chapter Dependencies

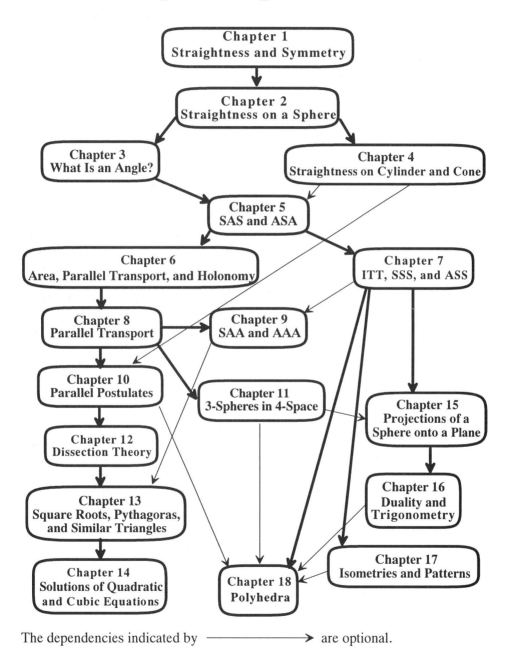

The dependencies indicated by ——————→ are optional.

How to Use This Book

Do not just pay attention to the words;
Instead pay attention to meanings behind the words.
But, do not just pay attention to meanings behind the words;
Instead pay attention to your deep experience of those meanings.
— Tenzin Gyatso, The Fourteenth Dalai Lama[†]

This quote expresses the philosophy upon which this book is based. Most of the chapters start intuitively so that they are accessible to a general reader with no particular mathematics background except imagination and a willingness to struggle with ones own experience of the meanings. However, the discussions in the book were written for mathematics majors and mathematics teachers and thus assume of the reader a corresponding level of interest and mathematical sophistication.

This book will present you with a series of problems. You should explore each question and write out your thinking in a way that can be shared with others. By doing this you will be able to actively develop ideas prior to passively reading or listening to comments of others. When working on the problems, you should be open-minded and flexible and let your thinking wander. Some problems will have short, fairly definitive answers, and others will lead into deep areas of meaning which can be probed almost indefinitely. You should not accept anything just because you remember it from school or because some authority says it's good. Insist on understanding (or seeing) why it is true or what it means for you. Pay attention to **your** deep experience of these meanings.

You should think about the problems and express your thinking about them even when you know you cannot do them completely. This is important because:

♦   It helps build self confidence.

♦   You will see what your real difficulties are.

♦   When you see a solution or proof later, then you will more likely see it as answering a question that you have.

[†] From an unpublished lecture in London, April, 1984. Used here by permission.

An important thing to keep in mind is that there is no one correct solution. There are many different ways of solving the problems — as many as there are ways of understanding the problems. *Insist on understanding* (or seeing) why it is true or what it means to you. Everyone understands things in a different way, and one person's "obvious" solution may not work for you. However, it is helpful to talk with others — listen to their ideas and confusions and then share your ideas and confusions with them.

Also, some of the problems are difficult to visualize in your head. Make models, draw pictures, use rubber bands on a ball, use scissors and paper — play!

For exploring properties on a sphere it is important that you have a model of a sphere that you can use. You can draw on worn tennis balls ("worn" because the fuzz can get in the way) and they are the right size for ordinary rubber bands to represent great circles. You may find useful clear plastic spheres in craft stores. Most any ball you have around will work — you can even use an orange and then eat it when you get hungry.

Message to the Reader

> In mathematics, as in any scientific research, we find two tendencies present. On the one hand, the tendency toward *abstraction* seeks to crystalize the *logical* relations inherent in the maze of material that is being studied, and to correlate the material in a systematic and orderly manner. On the other hand, the tendency toward *intuitive understanding* fosters a more immediate grasp of the objects one studies, a live *rapport* with them, so to speak, which stresses the concrete meaning of their relations.
>
> As to geometry, in particular, the abstract tendency has here led to the magnificent systematic theories of Algebraic Geometry, of Riemannian Geometry, and of Topology; these theories make extensive use of abstract reasoning and symbolic calculation in the sense of algebra. Notwithstanding this, it is still as true today as it ever was that intuitive understanding plays a major role in geometry. And such concrete intuition is of great value not only for the research worker, but also for anyone who wishes to study and appreciate the results of research in geometry.
>
> — David Hilbert[O: Hilbert, p. iii][†]

These words were written in 1934 by the "father of Formalism," David Hilbert in the Preface to *Geometry and the Imagination* by Hilbert and S. Cohn-Vossen. Hilbert has emphasized the point which I wish to make in this book:

> *Meaning is important in mathematics and geometry is an important source of that meaning.*

I believe that mathematics is a natural and deep part of human experience and that experiences of meaning in mathematics are accessible to everyone. Much of mathematics is not accessible through formal approaches except to those with specialized learning. However, through the use of non-formal experience and geometric imagery, many levels of meaning in mathematics can be opened up in a way that most human beings can experience and find intellectually challenging and stimulating.

Formalism contains the power of the meaning but not the meaning. It is necessary to bring the power back to the meaning.

[†]The information in the square brackets refers to the Bibliography. The initial letter indicates the section in the Bibliography and the name is the first author.

A proof as we normally conceive of it is not the goal of mathematics — it is a tool — a means to an end. The goal is understanding. Without understanding we will never be satisfied — with understanding we want to expand that understanding and to communicate it to others.

Many formal aspects of mathematics have now been mechanized and this mechanization is widely available on personal computers or even hand-held calculators, but the experience of meaning in mathematics is still a human enterprise that is necessary for creative work.

In this book I invite the reader to explore the basic ideas of geometry from a more mature standpoint. I will suggest some of the deeper meanings, larger contexts, and interrelations of the ideas. I am interested in conveying a different approach to mathematics, stimulating the reader to take a broader and deeper view of mathematics, and to experience for her- or himself a sense of mathematizing. Through an active participation with these ideas, including exploring and writing about them, people can gain a broader context and experience. This active particpation is vital for anyone who wishes to understand mathematics at a deeper level as well as vital for anyone wishing to understand something in their experience through the vehicle of mathematics.

This is particularly true for teachers or perspective teachers who are approaching related topics in the school curriculum. All too often we convey to students that mathematics is a closed system, with a single answer or approach to every problem, and often without a larger context. I believe that even where there are strict curricular constraints, there is room to change the meaning and the experience of the mathematics in the classroom.

Proof as Convincing Argument That Answers — Why?

Much of our view of the nature of mathematics is intertwined with our notion of what is a proof. This is often particularly true with geometry which has traditionally been taught in high school in the context of "two-column" proofs. The course materials in this book are based on a view of proof as a convincing argument that answers a why-question.

Why is $3 \times 2 = 2 \times 3$? To say, "It follows from the Commutative Law," does not answer the why-question. But most people will be convinced by, "I can count three 2's and then two 3's and see that they are both equal to the same six." OK, now why is $2,657,873 \times 92,564 = 92,564 \times 2,657,873$? We can not count this — it is too large. But is there a way to see $3 \times 2 = 2 \times 3$ without counting? Yes.

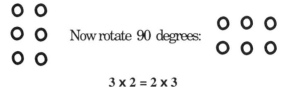

Figure 0.1 $3 \times 2 = 2 \times 3$

Most people will not have trouble extending this proof to include 2,657,873 x 92,564 = 92,564 x 2,657,873 or the more general $n \times m = m \times n$. Note that for the above to make sense I must have a meaning for 3×2 and a meaning for 2×3 and these meanings must be different. So naturally I have the question: "Why (or in what sense) are these meanings related?" A proof should help me experience relationships between the meanings. In my experience, to perform the formal mathematical induction proof starting from Peano's Axioms does not answer anyone's why-questions unless it is such a question as: "Why does the Commutative Law follow from Peano's Axioms?" Most people (other than logicians) have little interest in that question.

> CONCLUSION 1: *In order for me to be satisfied by a proof, the proof must answer my why-question and relate my meanings of the concepts involved.*

As further evidence toward this conclusion, you have probably had the experience of reading a proof and following each step logically but still not being satisfied because the proof did not lead you to experience the answer to your why-question. In fact most proofs in the literature are not written out in such a way that it is possible to follow each step in a logical formal way. Even if they were so written, most proofs would be too long and complicated for a person to check each step. Furthermore, even among mathematics researchers, a formal logical proof that they can follow step-by-step is not always satisfying. For example, my shortest research paper ["A simplicial complex whose product with any ANR is a simplicial complex," *General Topology.* **3** (1973), pp81-83] has a very concise simple proof that anyone who understands the terms involved can easily follow logically step-by-step. But, I have received more questions from other mathematicians about that paper than about any of my other research papers and most of the questions were of the sort: "Why is it true?" "Where did it come from?" "How did you see it?" They accepted the proof logically but were not satisfied.

Let us look at another example — the Vertical Angle Theorem: *If l and l' are straight lines, then the angle* α *is congruent to the angle* β.

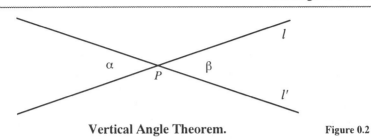

Vertical Angle Theorem.   Figure 0.2

As we will see in Problems 3 and 4 of this book, the proof of this theorem that is most convincing to someone depends on the meanings that person has of "angle" and "congruence." Some years ago, after I had been teaching this geometry course for over 10 years, several proofs that were convincing to me were presented by students in the class. But one student found the proofs not so convincing and offered a very straightforward simple proof of her own. My first reaction was that her argument could not possibly be a proof — it was too simple and did not involve everything in the standard proof. But she persisted patiently for several days and my meanings for angle congruence deepened. Now her proof is much more convincing to me than the standard one. I hope you will have similar experiences while working through this book.

CONCLUSION 2:   *A proof that satisfies someone else may not satisfy me because their meanings and why-questions are different from mine.*

You may ask: "But, at least in plane geometry, isn't an angle an angle? Don't we all agree on what an angle is?" Well, yes and no. Consider this acute angle:

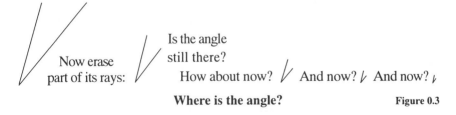

Where is the angle?   Figure 0.3

The angle is somehow *at the corner*, yet it is difficult to express this formally. As evidence, I looked in all the plane geometry books in the university library and found their definitions for "angle." I found nine different definitions! Each expressed a different meaning or aspect of "angle" and thus, each would potentially lead to a different proof of the Vertical Angle Theorem. We will see this more when we discuss Problems 3 and 4.

Sometimes we have legitimate why-questions even with respect to statements traditionally accepted as axioms. The Commutative Law above is one possible example. Another one is Side-Angle-Side (or SAS): *If two triangles have two sides and the included angle of one congruent to two sides and the included angle of the other, then the triangles are congruent.* You can find SAS listed in some geometry textbooks as an axiom to be assumed; in others it is listed as a theorem to be proved and in still others as a definition of the congruency of two triangles. But clearly one can ask, "Why is SAS true in the plane?" This is especially true since SAS is false for (geodesic) triangles on the sphere. So one can naturally ask, "Why is SAS true on the plane but not on the sphere?"

I have been teaching a geometry course based on the material in this book for a long time now (since 1974). One might expect that I have seen everything. But every year, about one-third of the students will show me a meaning or way of looking at the geometry that I have never thought of before and thus my own meaning and experience of geometry deepen. Looking back, I notice that these students who have shown me something new are mostly persons whose cultural backgrounds or race or gender are different from mine; and this is true even though most of the students in the class and I are white males.[†]

CONCLUSION 3: *Persons who are most different from me (for example, in terms of cultural background, race, gender) are most likely to have different meanings and thus have different why-questions and different proofs.*

You should check this out in your own experience. Note that this conclusion implies that I must listen particularly carefully to the meanings and proofs expressed by females and persons from other cultures and races because there is much which they see which I do not see. We should also be more critical of the standard histories of mathematics and mathematicians which have a decidedly Eurocentric emphasis.

COROLLARY: *I can learn much about mathematics by listening to persons whose cultural backgrounds, race, or gender are different than mine.*

As we personally experience Conclusions 1, 2 3, above, then we are led to the following conclusion.

CONCLUSION 4: *If I experience 1, 2, and 3, then other persons (for example, my students) are also likely to have similar experiences.*

[†]For details of this data and further discussion see "I Learn Mathematics From Persons Who Differ From Me" a paper in preparation by the author.

Chapter 1
Straightness and Symmetry

Straight is that of which the middle is in front of both extremities.
— Plato, *Parmenides,* 137 E [**A**: Plato]

A straight line is a line which lies evenly with the points on itself.
— Euclid, *Elements*, Definition 4 [**A**: Euclid]

Wisdom will save you from the ways of wicked men, from men whose words are perverse, who leave the straight paths to walk in the dark ways,...whose paths are crooked and who are devious in their ways.
— *The Holy Bible*, Proverbs 2:12-15 [**A**: Bible]

Verily, this is My Way,
Leading straight, follow it:
Follow not other paths:
They will scatter you about
From His great path:
— *Holy Qur-An*, Sura VI, verse 153 [**A**: Koran]

In keeping with the spirit of the approach to geometry discussed in the Preface and Chapter 0, we begin with a question that encourages you to explore deeply a concept that is fundamental to all that will follow. We ask you to build a notion of straightness for yourself at the beginning of the course rather than accept a certain number of assumptions about straightness. Though it is difficult to formalize, straightness is a natural human concept.

PROBLEM 1. *When Do You Call a Line Straight?*

Look to your experiences. It might help to think about how you would explain straightness to a 5-year-old (or how the 5-year-old might explain it to you!). If you use a "ruler," how do you know if the ruler is straight? How can you check it? What properties do straight lines have that distinguish them from non-straight lines?

Think about the question in four related ways:

1

1. How can you check in a practical way if something is straight — without assuming that you have a ruler, for then we will ask, "How can you check that the ruler is straight?"

2. How do you construct something straight — lay out fence posts in a straight line, or draw a straight line?

3. What symmetries does a straight line have? A symmetry of a geometric figure is a reflection, rotation, translation, or composition of them which preserves the figure. For example, the letter "T" has reflection symmetry about a vertical line through its middle, and the letter "Z" has rotation symmetry if you rotate it half a revolution about its center.

4. Can you write a definition of "straight line"?

Suggestions

Look at your experience. At first, you will look for examples of physical world or natural straightness. Go out and actually try walking along a straight line and then along a curved path; try drawing a straight line and checking that a line already drawn is straight.

Look for things that you call "straight." Where do you see straight lines? Why do you say they are straight? Look for both physical lines and non-physical uses of the word "straight."

You are likely to bring up many ideas of straightness. It is necessary then to think about what is common among all of these straight phenomena.

As you look for properties of straight lines that distinguish them from non-straight lines, you will probably remember the following statement (which is often taken as a definition in high school geometry): *A line is the shortest distance between two points.* But can you ever measure the lengths of all the paths between two points? How do you find the shortest path? If the shortest path between two points is in fact a straight line, then is the converse true? Is a straight line between two points always the shortest path? We will return to these questions in later chapters.

A powerful approach to this problem is to think about lines in terms of symmetry. This will become increasingly important as we go on to other surfaces (spheres, cones, cylinders, etc.) Two symmetries of lines are:

◆   reflection symmetry in the line, also called bilateral symmetry — reflecting (or mirroring) an object over the line.

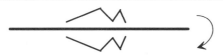

Reflection symmetry of a straight line.   Figure 1.1

◆   Half-turn symmetry — rotating 180° about any point on
the line.

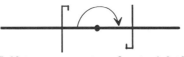

Half-turn symmetry of a straight line.   Figure 1.2

Although we are focusing on a symmetry of the line in each of these
examples, notice that the symmetry is not a property of the line by itself
but includes the line and the space around the line. The symmetries preserve the local environment of the line. Notice how in reflection and half-
turn symmetry the line and its local environment are both part of the
symmetry action and how the relationship between them is integral to the
action. In fact, reflection in the line does not move the line at all but ex-
hibits a way in which the space on the two sides of the line are the same.
A *symmetry of a figure* is a transformation of a region of space that preserves distances and angle measures and which takes the part of figure in
the region onto (possibly another part of) itself.

   Try to think of other symmetries as well (there are quite a few).
Some symmetries hold only for straight lines, while some work with
other curves too. Try to determine which ones are specific to straight
lines and why. Also think of practical applications of these symmetries
for constructing a straight line or for determining if a line is straight.

How Do You Construct a Straight Line?

   As for how to construct a straight line, one method is simply to fold
a piece of paper; the edges of the paper needn't even be straight. This
utilizes symmetry (can you see which one?) to produce the straight line.
Carpenters also use symmetry to determine straightness — they put two
boards face to face, plane the edges until they look straight, and then turn
one board over so the planed edges are touching. See Figure 1.3. They
then hold the boards up to the light. If the edges are not straight, there
will be gaps between the boards which light will shine through.

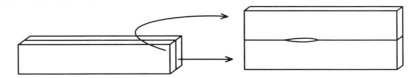

Figure 1.3   **Carpenter's method for checking straightness.**

When grinding an extremely accurate flat mirror, the following technique is sometimes used: Take three approximately flat pieces of glass and put pumice between the first and second pieces and grind them together. Then do the same for the second and the third pieces and then for the third and first pieces. Repeat many times and all three pieces of glass will become very accurately flat. See Figure 1.4. Do you see why this works? What does this have to do with straightness?

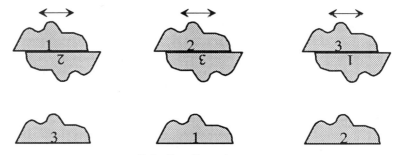

Figure 1.4   **Grinding flat mirrors.**

Imagine (or actually do it!) walking while pulling a long silk thread with a small stone attached. When will the stone follow along your path? Why? This property is used to pick up a fallen water skier. The boat travels by the skier along a straight line and thus the tow rope follows the path of the boat. Then the boat turns in an arc in front of the skier. Since the boat is no longer following a straight path, the tow rope will move in toward the fallen skier.

Another idea to keep in mind is that straightness must be thought of as a local property. Part of a line can be straight even though the whole line may not be. For example, if we agree that this line is straight,

and then we add a squiggly part on the end, like this:

then would we now say that the original part of the line is not straight, even though it hasn't changed, only been added to? Also note that we are not making any distinction here between "line" and "line segment." The more generic term "line" generally works well to refer to any and all lines and line segments, both straight and non-straight.

Now look at symmetries. What symmetries does a straight line have? How do they fit with the examples that you have found and those mentioned above? Can we use any of the symmetries of a line to define straightness?

Think about and formulate some answers for these questions before you read any further. You are the one laying down the definitions. Do not take anything for granted unless you see why it is true. No answers are predetermined. You may come up with something that we have never imagined. Consequently, it is important that you persist in following your own ideas. Reread the section "How to Use This Book" on pages xviii and xix.

This icon will be used throughout the book to indicate places where you should pause and not read further until you have expressed your thinking and ideas through writing or through talking to someone else.

The Symmetries of a Line

Reflection-in-the-line symmetry: It is most useful to think of reflection as a "mirror" action with the line as an axis rather than as a "flip-over" action which involves an action in 3-space. In this way one can extend the notion of reflection symmetry to a sphere (the flip-over action is not possible on a sphere). Notice that this symmetry cannot be used as a definition for straightness because we use straightness to define reflection symmetry — the definition would be circular.

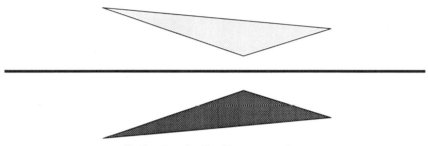

Figure 1.5   **Reflection-in-the-line symmetry.**

♦   *Practical application:* One can produce a straight line by folding a piece of paper because this action forces reflection-in-itself symmetry along the crease. Above we showed a carpenter's example.

Reflection-perpendicular-to-the-line symmetry: A   reflection through **any** axis perpendicular to the line will take the line onto itself. Note that circles also have this symmetry about any diameter.

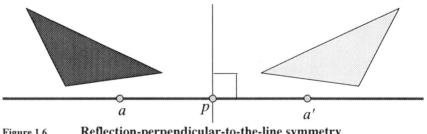

Figure 1.6   **Reflection-perpendicular-to-the-line symmetry.**

♦   *Practical applications:* You call tell if a straight segment is perpendicular to a mirror by seeing if it looks straight with its reflection. Also, a straight line can be folded onto itself.

Half-turn symmetry: A rotation through a half of a full revolution about any point *p* on the line takes the part of the line before *p* onto the part of the line after *p* and vice versa. Note that some non-straight lines, such as the letter "Z" also have half-turn symmetry.

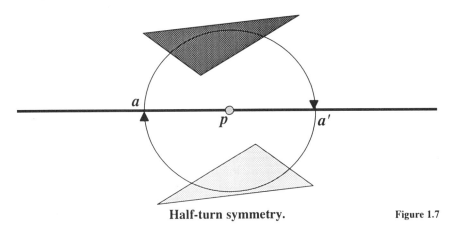

Half-turn symmetry.   Figure 1.7

◆ *Practical applications*: This is the principle behind hinges. If this is applied in 3-space, then you get a universal joint (a hinge that can fold in any direction).

Rigid-motion-along-itself symmetry: For straight lines in the plane, we call this *translation symmetry*. Any portion of a straight line may be moved along the line without leaving the line. This property of being able to move rigidly along itself is not unique to straight lines; circles (rotation symmetry) and circular helixes (screw symmetry) have this property as well.

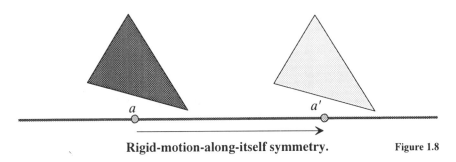

Rigid-motion-along-itself symmetry.   Figure 1.8

◆ *Practical applications*: Slide joints such as in trombones, drawers, nuts and bolts, pulleys, etc. all utilize this symmetry.

3-dimensional rotation symmetry: In a 3-dimensional space, rotate the line around itself through *any* angle using itself as an axis.

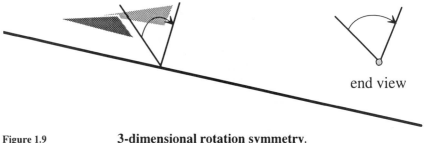

end view

Figure 1.9   **3-dimensional rotation symmetry.**

◆ *Practical applications*: This symmetry can be used to check the straightness of any long thin object such as a stick by twirling the stick with itself as the axis. If the stick does not appear to wobble, then it is straight. This is used for pool cues, axles, etc.

Central symmetry, or point symmetry: Central symmetry through the point *p* sends any point *a* to the point at the same distance from *p* but on the diametrically opposite side. In two dimensions central symmetry does not differ from half-turn symmetry in its end result.

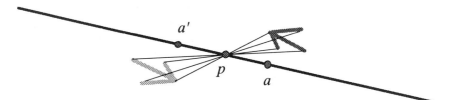

Figure 1.10   **Central symmetry.**

◆ In 3-space, central symmetry produces a result different than any single rotation or reflection (though one can check that it does give the same result as the composition of three reflections through mutually perpendicular planes). To experience central symmetry in 3-space, hold your hands in front of you with the palms facing each other and your left thumb up and your right thumb down. Your two hands now have approximate central symmetry about a point midway between the center of the palms.

Similarity symmetry, or self-similarity: Any segment of a straight line (and its environs) is similar to (i.e., can be magnified or shrunk to become the same as) any other segment.

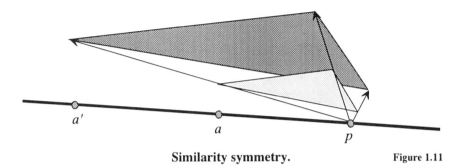

Similarity symmetry.   Figure 1.11

◆   Logarithmic spirals like the chambered nautilus have self-similarity as do many fractals.

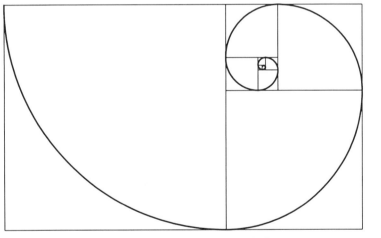

Golden ratio logarithmic spiral.   Figure 1.12

Clearly, other objects besides lines have some of the symmetries mentioned here. It is important for you to construct your own such examples and for you to attempt to find an object that has all of the symmetries but is not a line. This will help you to experience that straightness and the seven symmetries discussed here are intimately related. You should come to the conclusion that while other curves and figures have some of these symmetries, only the straight line has all of them.

Returning to one of the original questions, how would we construct a straight line? One way would be to use a "straight edge" — something that we accept as straight. Notice that this is different from the way that we would draw a circle. When using a compass to draw a circle, we are not starting with a figure that we accept as circular; instead, we are using a fundamental property of circles that the points on a circle are a fixed distance from the center. Can we use the symmetry properties of a straight line to construct a straight line? Is there a tool (serving the role of a compass) which will draw a straight line? For an interesting discussion of this question see *How to Draw a Straight Line: A Lecture on Linkages* by A.B. Kempe [**Q**: Kempe] which shows the following apparatus:

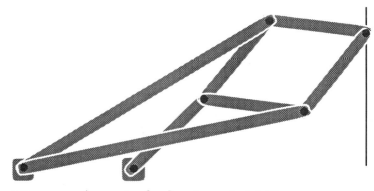

Figure 1.13   **Apparatus for drawing a straight line.**

Local (and Infinitesimal) Straightness

Previously, you saw how a straight line has reflection-in-the-line symmetry and half-turn symmetry: One side of the line is the same as the other. But, as pointed out above, straightness is a local property in that whether a segment of a line is straight depends only on what is near the segment and does not depend on anything happening away from the line. Thus each of the symmetries must be able to be thought of (and experienced) as applying only locally. This will become particularly important later when we investigate straightness on the cone and cylinder. (See the discussions in Chapter 4.) For now, it can be experienced in the following way:

> *When a piece of paper is folded not in the center, the crease is still straight even though the two sides of the crease on the paper are not the same. So what is the role of the sides when*

*we are checking for straightness using reflection symmetry?
Think about what is important near the crease in order to have
reflection symmetry.*

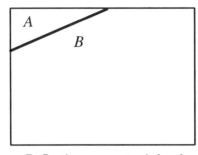

Reflection symmetry is local.   Figure 1.14

When we talk about straightness as a local property, you may bring
out some notions of scale. For example, if one sees only a small portion
of a very large circle, it will be indistinguishable from a straight line.
This can be experienced easily on many of the modern graphing pro-
grams for computers. Also a microscope with a zoom lens will provide
an experience of zooming. If a curve is smooth (or differentiable), then if
one "zooms in" on any point of the curve, eventually the curve will be indistinguishable from a straight line segment.

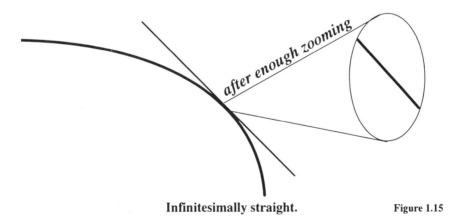

Infinitesimally straight.   Figure 1.15

This property is called *infinitesimally straight*, or in more standard
terminology, *differentiable*. We say that a curve is *infinitesimally straight*
at a point *p* if there is a straight line *l* such that if we zoom in enough

on p, the line and the curve become indistinguishable.[†] When the curve is parametrized by arc length it is equivalent to the curve having a well-defined velocity vector at each point.

In contrast, we can say that a curve is *locally straight at a point* if that point has a neighborhood that is straight. In the physical world the usual use of both *smooth* and *locally straight* are dependent on the scale at which they are viewed. For example, we may look at an arch made out of wood that at a distance appears as a smooth curve; then as we move in closer we see that the curve is made by many short straight pieces of finished (planed) boards, but when we are close enough to touch it, we see that its surface is made up of smooth waves or ripples, and under a microscope we see the non-smoothness of numerous twisting fibers. See Figure 1.16.

[†]This is equivalent to the usual definitions of being differentiable at p. For example, if $t(x) = f(p) + f'(p)(x-p)$ is the equation of the line tangent to the curve $(x, f(x))$ at the point $(p, f(p))$, then, given $\varepsilon > 0$ (the distance of indistinguishability), there is a $\delta > 0$ (the radius of the zoom window) such that, for $|x - p| < \delta$ (for x within the zoom window), $|f(x) - t(x)| < \varepsilon$ ($f(x)$ is indistinguishable from $t(x)$). This last inequality may look more familiar in the form:

$$f(x) - t(x) = f(x) - f(p) - f'(p)(x-p) = \{ [f(x) - f(p)]/(x-p) - f'(p) \}(x-p) < \varepsilon.$$

In general, the value of δ might depend on p as well as on ε. Often the term *smooth* is used to mean continuously differentiable which the interested reader can check is equivalent (on closed finite intervals) to, for each $\varepsilon > 0$, there being one $\delta > 0$ that works for all p.

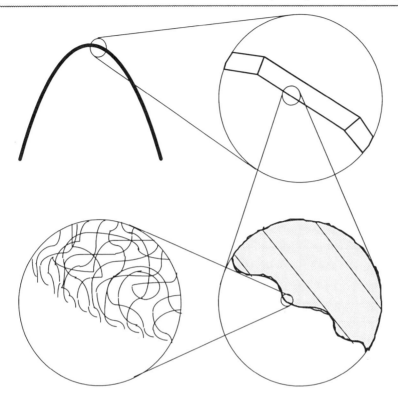

Straightness and smoothness depend on the scale.   Figure 1.16

Chapter 2
Straightness on a Sphere

> ... it will readily be seen how much space lies between the two places themselves on the circumference of the large circle which is drawn through them around the earth. ... [W]e grant that it has been demonstrated by mathematics that the surface of the land and water is in its entirety a sphere, ... and that any plane which passes through the center makes at its surface, that is, at the surface of the earth and of the sky, great circles, and that the angles of the planes, which angles are at the center, cut the circumferences of the circles which they intercept proportionately, ...
> — Ptolemy, *Geographia* (ca. 150 AD) Book One, Chapter II

Drawing upon the intuitive ideas about straightness developed in the first problem, Problem 2 asks for a criterion for straightness on a sphere. It is important for you to see that if you are not building a notion of straightness for yourself (for example, if you are taking ideas from books without thinking deeply about them), then you will have difficulty building a concept of straightness on a sphere. Only by developing a personal meaning of straightness for oneself does it become part of one's active intuition. We say *active* intuition to emphasize that intuition is in a process of constant change and enrichment, that it is not static.

PROBLEM 2. *What Is Straight on a Sphere?*

Imagine yourself to be a bug crawling around on a sphere. (This bug can neither fly nor burrow into the sphere.) The bug's universe is just the surface; it never leaves it. What is "straight" for this bug? What will the bug see or experience as straight? How can you convince yourself of this? Use the properties of straightness (such as symmetries) which you talked about in Problem 1.

Show (i.e., convince yourself, and give an argument to convince others) *that the great circles on a sphere are straight with respect to the sphere, and that no other circles on the sphere are straight with respect to the sphere.*

Suggestions

Great circles are those circles which are the intersection of the sphere with a plane through the center of the sphere. Examples: Any longitude line and the equator are great circles on the earth — consider any pair of opposite points as being the poles and thus the equator and longitudes with respect to any pair of opposite points will be great circles. See examples illustrated in Figure 2.1.

Great Circles.   **Figure 2.1**

The first step to understanding this problem is to convince yourself that great circles are straight lines on a sphere. Think what it is about the great circles that would make the bug experience them as straight. To better visualize what is happening on a sphere (or any other surface, for that matter), **you must use models**. This is a point we cannot stress enough. The use of models will become increasingly important in later problems, especially those involving more than one line. You must make lines on a sphere to fully understand what is straight and why. An orange or an old, worn tennis ball work well as spheres, and rubber bands make good lines. Also, you can use ribbon or strips of paper. Try placing these items on the sphere along different curves to see what happens.

Also look at the symmetries from Problem 1 to see if they hold for straight lines on the sphere. The important thing to remember here is to **think in terms of the surface of the sphere, not in 3-space**. Always try to imagine how things would look from the bug's point of view. A good example of how this type of thinking works is to look at an insect called a water strider. The water strider walks on the surface of a pond and has a very 2-dimensional perception of the world around it—to the water strider, there is no up or down; its whole world consists of the plane of the water. The water strider is very sensitive to motion and vibration on the water's surface, but it can be approached from above or below without its knowledge. Birds and fish take advantage of this fact. This is the type of thinking needed to adequately visualize properties of straight lines on the sphere.

Lines which are straight on a sphere (or other surfaces) are often called *geodesics*. This leads us to consider the concept of intrinsic or geodesic curvature versus extrinsic curvature. As an outside observer looking at the sphere in 3-space, all paths on the sphere, even the great circles, are curved — that is, they exhibit *extrinsic* curvature. But relative to the surface of the sphere (*intrinsically*), the lines may be straight. Be sure to understand this difference and to see why all symmetries (such as reflections) must be carried out intrinsically, or from the bug's point of view.

It is natural for you to have some difficulty experiencing straight on surfaces other than the plane and that consequently you will start to look at the properties of spheres and at the curves on spheres as 3-D objects. Imagining that you are a 2-dimensional bug walking on a sphere emphasizes the importance of experiencing straightness and will help you to shed your limiting extrinsic 3-D vision of the curves on a sphere. Ask yourself:

◆   What does the bug have to do, when walking on a non-planar surface, in order to walk in a straight line?

◆   How can the bug check if it is going straight?

Experimentation with models plays an important role here. Working with models that *you create* helps you to experience that great circles are, in fact, the only straight lines on the surface of a sphere. Convincing yourself of this notion will involve recognizing that straightness on the plane and straightness on a sphere have common elements. When you are comfortable with "great-circle-straightness," you will be ready to transfer the symmetries of straight lines on the plane to great circles on a sphere and, later, to geodesics on other surfaces. Here are some activities that you can try, or visualize, to help you experience great circles and their intrinsic straightness on a sphere. However, it is better for you to come up with your own experiences.

◆   Stretch something elastic on a sphere. It will stay in place on a great circle, but it will not stay on a small circle if the sphere is slippery. Here, the elastic follows a path that is approximately the shortest since a stretched elastic always moves so that it will be shorter. Using the shortest distance criterion directly is not a good way to check for straightness because one cannot possibly measure all paths. But, it serves a good purpose here.

◆   Roll a ball on a straight chalk line (or straight on a freshly painted floor!). The chalk (or paint) will mark the line of contact on the sphere and it will be a great circle.

◆   Take a stiff ribbon or strip of paper that does not stretch, and lay it "flat" on a sphere. It will only lie properly along a great circle. Do you see how this property is related to local symmetry? This is sometimes called the *Ribbon Test*. (See Appendix.)

◆   The feeling of turning and "non-turning" comes up. Why is it that on a great circle there is no turning and on a latitude line there is turning? Physically, in order to avoid turning, the bug has to move its left feet the same distance as its right feet. On a non-great circle (for example, a latitude line that is not the equator), the bug has to walk faster with the legs that are on the side closest to the equator. This same idea can be experienced by taking a small toy car with its wheels fixed so that, on a plane, it rolls along a straight line. Then on the sphere the car will roll around a great circle but it will not roll around other curves.

◆   Also notice that, on a sphere, straight lines are circles (points on the surface a fixed distance away from a given point) — special circles whose circumferences are straight! Note that the equator is a circle with two centers: the north pole and the south pole. In fact, any circle on a sphere has two centers.

These activities will provide you with an opportunity to investigate the relationships between a sphere and the geodesics of that sphere. Along the way, your experiences should help you to discover how great circles on a sphere have most of the same symmetries as straight lines on a plane.

You should pause and not read further until you have expressed your thinking and ideas about this problem.

Symmetries of Great Circles

Reflection-thru-itself symmetry: We can see this globally by placing a hemisphere on a flat mirror. The image in the mirror exactly recreates the hemisphere. Figure 2.2 shows a reflection through the great circle *g*.

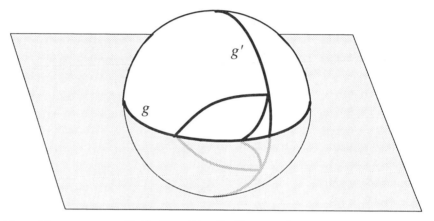

Figure 2.2   **Reflection-thru-itself symmetry.**

Reflection-perpendicular-to-itself symmetry: A reflection through any great circle will take any great circle (for example, *g′* in Figure 2.2) which is perpendicular to the original great circle onto itself.

Half-turn symmetry: A rotation through half of a full revolution about any point *p* on a great circle interchanges the part of the great circle on one side of *p* with the part on the other side of *p*.

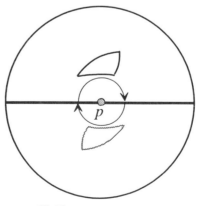

Figure 2.3   **Half-turn symmetry.**

Rigid-motion-along-itself symmetry: For great circles on a sphere we call this a translation along the great circle and a rotation around the poles of that great circle. This property of being able to move rigidly along itself is not unique to great circles since any circle on the sphere whose center is a pole of the great circle will also have the same symmetry.

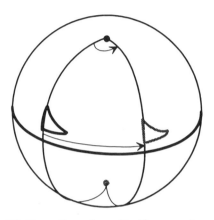

Rigid-motion-along-itself symmetry.   **Figure 2.4**

Central symmetry, or point symmetry: Viewed intrinsically (i.e., from the 2-dimensional bug's point-of-view), central symmetry through a point *p* on the sphere sends any point *a* to the point at the same great circle distance from *p* but on the opposite side.

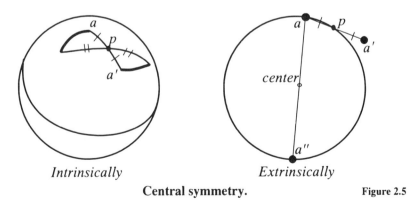

Intrinsically   *Extrinsically*
Central symmetry.   **Figure 2.5**

Extrinsically (i.e., from our 3-dimensional point-of-view), the only central symmetry of great circles is through the center of the sphere. (See

Figure 2.5.) Intrinsically, as on a plane, central symmetry does not differ from half-turn symmetry with respect to the end result. This distinction between intrinsic and extrinsic is important to experience at this point.

3-dimensional rotation symmetry: **This symmetry does not hold for great circles in 3-space**; however, it does hold for great circles in a 3-sphere. See Chapter 11.

Similarity symmetry, or self-similarity: **This symmetry does not hold on spheres**, as we shall see in Problem 19.

You will probably notice that other objects, besides great circles, have some of the symmetries mentioned here. It is important for you to construct such examples, and to attempt to find an object that has all of the symmetries mentioned here but is not a great circle. This will help you to realize that straightness and the five symmetries discussed here are intimately related.

Relationships with Differential Geometry

You have tried wrapping the sphere with a ribbon and noticed that the ribbon will only lie flat along a great circle. (If you haven't experienced this yet, then do it now before you go on.) Arcs of great circles are the only paths of a sphere's surface that are tangent to a straight line on a piece of paper wrapped around the sphere. If you wrap a piece of paper tangent to the sphere around a latitude circle (see Figure 2.6), then, extrinsically, the paper will form a portion of a cone and the curve on the paper will be an arc of a circle.

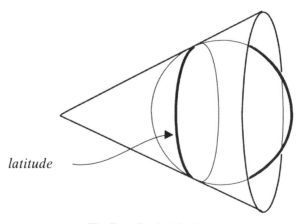

latitude

Figure 2.6   **Finding the intrinsic curvature.**

The *intrinsic curvature* of a path on the surface of a sphere can be defined as the curvature that one gets when one "unwraps" the path onto a plane. For related information see the Appendix, "A Geometric Introduction to Differential Geometry".

In differential geometry, mathematicians often talk about straight paths (geodesics) in terms of the velocity vector of the motion as one travels at a constant speed along that path. (The velocity vector is tangent to the curve along which the bug walks.) For example, as you walk along a great circle, the velocity vector to the circle changes direction, extrinsically, in 3-space where the change in direction is toward the center of the sphere. "Toward the center" is not a direction that makes sense to a 2-dimensional bug whose whole universe is the surface of the sphere. Thus, the bug does not experience the velocity vectors at points along the great circle as changing direction. In differential geometry, the rate of change, from the bug's point of view, is called the *covariant* (or *intrinsic*) *derivative*. As the bug traverses a geodesic, the covariant derivative of the velocity vector is zero. This can also be expressed in terms of *parallel transport* which is discussed in Chapters 6 and 8. See also the Appendix.

Chapter 3
What Is an Angle?

A **plane angle** is the inclination to one another of two lines in a
plane, which meet one another, but do not lie in a straight line.
— Euclid, *Elements*, Definition 8 [A: Euclid]

In Problems 3 and 4 you will be thinking about angles. It is not always necessary to do the problems in order, and in this case, you may
find it easier to do Problem 4 before Problem 3. It may help to think
about the properties of angles before trying to prove theorems about
them. In a sense, you should be working on both Problems 3 and 4 at the
same time because they are so closely intertwined.

PROBLEM 3. *Vertical Angle Theorem (VAT)*

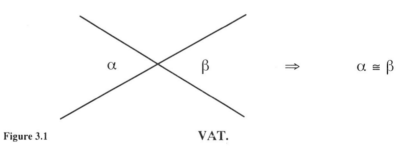

Figure 3.1   **VAT.**

*Prove: A pair of opposite angles formed by two intersecting
straight lines are congruent.* [Note: Angles such as α and β
are called **vertical angles**.]

*We do not have in mind a formal two-column proof as used
in high school geometry. Mathematicians in actual practice
usually use a proof to mean a discussion sufficient to convince
any reasonable skeptic.* [Hint: Show how you would "move" α
to make it coincide with β. What symmetry properties of
straight lines are you using?]

Does your proof also work on a sphere?

PROBLEM 4. *What Is an Angle?*

Give some possible definitions of the term "Angle." Do all of these definitions apply to both the plane and the sphere? What are the advantages and disadvantages of each?

What does it mean for two angles to be congruent? How can we check?

Suggestions

Textbooks usually give some variant of the definition: *An angle is the union of two rays (or segments) with a common endpoint.* If we start with two straight line segments with a common endpoint, and then add squiggly parts onto the ends of each one, would we say that the angle has changed as a result? Likewise, look at the angle formed at the lower-left-hand corner of this piece of paper. Even first grade students will recognize this as an example of an angle. Now, tear off the corner (at least in your imagination). Is the angle still there, at the corner you tore off? Now tear away more of the sides of the angles, being careful not to tear through the corner.

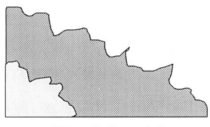

Where is the angle?   Figure 3.2

The angle is still there at the corner, isn't it? So what part of the angle determines how large the angle is, or if it is an angle at all? What is the angle? It seems it can not be merely a union of two rays. Here is one of the many cases where children seem to know more than we do. Paying attention to these insights, can we get better definitions of angle? Do not expect to find one formal definition that is completely satisfactory; it seems likely that no formal definition can capture all aspects of our experience of what an angle is.

Symmetries were an important element of your solutions for Problems 1 and 2. They will be very useful for this problem, as well. It is perfectly valid to think about measuring angles in this problem, but proofs utilizing line symmetries are generally simpler. It often helps to think of the vertical angles as whole geometric figures. Also, keep in mind that there are many different ways of looking at angles, so there are many ways of proving the vertical angle theorem. Make sure that your notions of angle and angle congruency in Problem 4 are consistent with your proofs in Problem 3, and vice versa.

There are at least three different perspectives from which one can define "angle."

♦   a *dynamic* notion of angle — angle as movement;

♦   angle as *measure*; and,

♦   angle as *geometric shape*.

Each of them, separately or together, might help you prove the Vertical Angle Theorem. A *dynamic* notion of angle involves an action: a rotation, a turning point, or a change in direction between two lines. Angle as *measure* may be thought of as the arc length of a circular sector or the ratio between areas of circular sectors. Thought of as a *geometric shape,* an angle may be seen as the delineation of space by two intersecting lines. Each of these perspectives carries with it methods for checking angle congruency. You can check the congruency of two dynamic angles by verifying that the actions involved in creating or replicating them are the same. If you feel that an angle is a measure, then you must verify that both angles have the same measure. If you describe angles as geometric shapes, then one angle should be made to coincide with the other using isometries in order to prove angle congruence. Which of the above definitions has the most meaning for you? Are there any other useful ways of describing angles?

Note that we sometimes talk about ***directed angles***, or angles with direction. When considered as directed angles, we say that the angles α and β in Figure 3.3 are not the same but have equal magnitude and opposite directions (or sense). Note the similarity to the relationship between line segments and vectors.

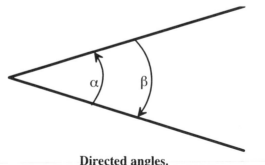

Figure 3.3   **Directed angles.**

You should pause and not read further until you have expressed your own thinking and ideas about Problems 3 and 4.

Hints for Three Different Proofs

In the following section, we will give hints for three different proofs of the Vertical Angle Theorem. Note that a particular notion of angle is assumed in each proof. Pick one of the proofs, or find your own different proof that is consistent with a notion of angle and angle congruence that is most meaningful to you.

1st proof:

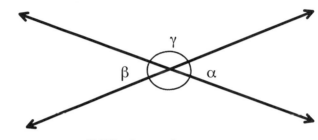

Figure 3.4   **VAT using angle as measure.**

Each line is a 180° angle. Thus, $\alpha + \gamma = \beta + \gamma$.

Therefore we can conclude that $\alpha \cong \beta$. But why is this so? Is it always true that if one subtracts a given angle from two 180° angles then the remaining angles are congruent?

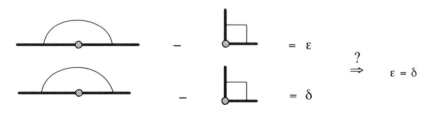

Figure 3.5   **Subtracting angles and measures.**

Numerically, it does not make any difference how one subtracts an angle, but geometrically it makes a big difference. Behold Figure 3.6! Here, ε really cannot be considered the same as δ. Thus, measure does not completely express what we *see* in the geometry of this situation. If you wish to salvage this notion of angle as measure, then you must explain *why* it is that in this proof of the Vertical Angle Theorem γ can be subtracted from both sides of the equation $\alpha + \gamma = \beta + \gamma$.

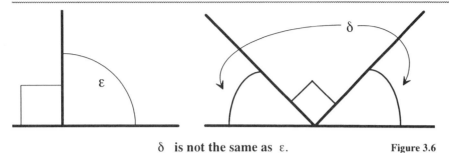

δ **is not the same as** ϵ.   **Figure 3.6**

2nd proof: Consider two overlapping lines and choose any point on them. Rotate one of the lines, maintaining the point of intersection and making sure that the other line remains fixed.

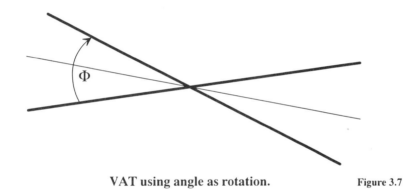

VAT using angle as rotation.   **Figure 3.7**

What happens? What notion of angle and angle congruency is at work here?

3rd proof: What symmetries will take α onto β? See Figure 3.1 or 3.4.

Chapter 4
Straightness on Cylinder and Cone

> Definition: When a straight line set up on a straight line, makes the adjacent angles equal to one another, each of the equal angles is **right**, and the straight line standing on the other is called a **perpendicular** to that on which it stands.
>
> Let the following be postulated: That all right angles are equal to one another.
>
> — Euclid, *Elements,* Postulate 4 [**A**: Euclid]

It is time to return to straightness, but now the goal is to think intrinsically. By the end of Problem 5, you should be comfortable with straightness as a *local intrinsic notion* — this is the bug's view. This notion of straightness is the basis for the notion of *geodesics* in differential geometry.

When looking at great circles on the surface of a sphere, we were able (except in the case of central symmetry) to see all the symmetries of straight lines from global extrinsic points of view. For example, a great circle extrinsically divides a sphere into two hemispheres which are mirror images of each other. When looking at straightness on a sphere, it is a natural tendency to use the more familiar and comfortable extrinsic lens instead of taking the bug's local and intrinsic point of view. However, on a cone and cylinder you must use the local, intrinsic point of view because there is no extrinsic view that will work.

In Problem 2, you were asked to consider straightness on a sphere from the bug's point of view. Problem 5 also encourages you to look at straightness from the bug's point of view. The questions in Problem 5 are similar to those in Problem 2, but this time the surfaces are the cylinder and a cone.

PROBLEM 5. Intrinsic Straight Lines on Cones and Cylinders

What lines are straight with respect to the surface of a cone or a cylinder? Why? Why not?

28

Suggestions

Make paper models, but consider the cone or cylinder as continuing indefinitely with no top or bottom (except, of course, at the conepoint). Again, imagine yourself as a bug whose whole universe is a cone or cylinder. As the bug crawls around on one of these surfaces, what will the bug experience as straight? As with the sphere, paths which are straight with respect to a surface are often called the "geodesics" for the surface.

As you begin to explore these questions, it is likely that many other related geometric ideas will arise. Do not let seemingly irrelevant excess geometric baggage worry you. Often, you will find yourself getting lost in a tangential idea, and that's understandable. Ultimately, however, the exploration of related ideas will give you a richer understanding of the scope and depth of the problem. In order to work through possible confusion on this problem, try some of the following suggestions which others have found helpful. Each suggestion involves constructing or using models of cones and cylinders:

◆   If we make a cone or cylinder by rolling up a sheet of paper, will "straight" stay the same for the bug when we unroll it? Conversely, if we have a straight line drawn on a sheet of paper and roll it up, will it continue to be experienced as straight for the bug crawling on the paper?

◆   Lay a stiff ribbon or straight strip of paper on a cylinder or cone. Convince yourself that it will follow a straight line with respect to the surface. Also, convince yourself that straight lines on the cylinder or cone, when looked at locally and intrinsically, have the same symmetries as on the plane.

◆   If you intersect a cylinder by a flat plane and unroll it, what kind of curve do you get? Is it ever straight?

◆   On a cylinder or cone, can a geodesic ever intersect itself? How many times?

◆   Can there be more than one geodesic joining two points on a cylinder or cone? How many? Is there always at least one?

◆   You may find it helpful to explore cylinders first before beginning to explore cones. This problem has many aspects, but focusing at first on the cylinder will simplify some things.

There are several important things to keep in mind while working on this problem. First, **you absolutely must make models**. If you attempt to visualize lines on a cone or cylinder, you are bound to make claims that you would easily see are mistaken if you investigated them on an actual cone or cylinder. Many students find it helpful to make models using transparencies.

Second, as with the sphere, you must think about lines and triangles on the cone and cylinder in an intrinsic way — always look at things from a bug's point of view. We are not interested in what's happening in 3-space; only what you would see and experience if you were restricted to the surface of a cone or cylinder.

And last, but certainly not least, you must look at cones of different shapes, i.e., cones with varying cone angles.

Cones with Varying Cone Angles

Geodesics behave differently on differently shaped cones. So an important variable is the cone angle. The *cone angle* is generally defined as the angle measured around the point of the cone on the surface. Notice that this is an intrinsic description of angle. The bug could measure a cone angle, first, by making a model of a one-degree angle and, then, by determining how many of them it would take to go around the cone point. We can determine the cone angle extrinsically in the following way: If we cut the cone along a generator and flatten it, then the measure of the cone angle is the number of degrees in the planar sector.

For example, if we take a piece of paper and bend it so that half of one side meets up with the other half of the same side, we will have a 180-degree cone:

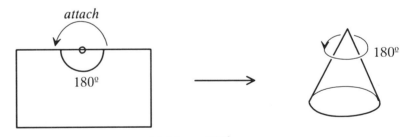

Figure 4.1   **Making a 180° cone.**

A 90º cone is also easy to make — just use the corner of a sheet of paper and bring one side around to meet with the adjacent side. Also be sure to look at larger cones. One convenient way to do this is to make a cone with a variable cone angle. This can be accomplished by taking a sheet of paper and cutting (or tearing) a slit from one edge to the center. (See Figure 4.2.) A rectangular sheet will work but a circular sheet is easier to picture. Note that it is not necessary that the slit be straight!

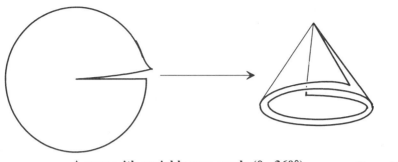

A cone with variable cone angle (0 - 360°).   Figure 4.2

You've already looked at a 360º cone in some detail — it's just a plane. The cone angle can also be larger than 360º. A common larger cone is the 450º cone. You probably have a cone like this somewhere on the walls, floor, and ceiling of your room. You can easily make one by cutting a slit in a piece of paper and inserting a 90º slice (360º + 90º = 450º):

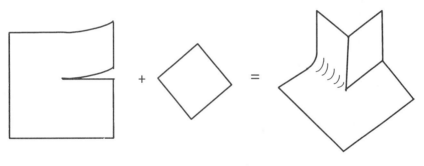

How to make a 450° cone.   Figure 4.3

You may have trouble believing that this is a cone, but remember that just because it can't hold ice cream, doesn't mean it's not a cone. If the folds and creases bother you, they can be taken out — the cone will look ruffled instead. It is important to realize that when you change the shape of the cone like this (i.e., by ruffling), you are only changing its extrinsic appearance. Intrinsically (from the bug's point of view) there is no difference. You can even ruffle the cone so that it will hold ice cream if you like, although changing the extrinsic shape in this way is not useful to a study of its intrinsic behavior.

You can also make a cone with variable angle of more than 180° by taking two sheets of paper and slitting them together to their centers as in

Figure 4.4. Then tape the left side of the top slit to the right side of the bottom slit as pictured.

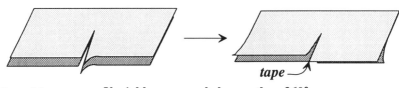

Figure 4.4   **Variable cone angle larger than 360°.**

It may be helpful for you to discuss some definitions of a cone. The following is one definition: *Take any simple (non-intersecting) closed curve* ***a*** *on a sphere and consider a point* ***P*** *at the center of the sphere. A* ***cone*** *is the union of the rays that start at* ***P*** *and go through each point on* ***a***. The cone angle is then equal to (length of ***a***)/(radius of sphere), in radians. Do you see why?

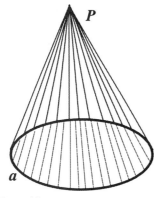
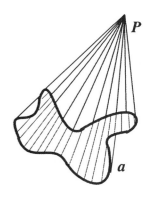

Figure 4.5   **Cones.**

Experiment by making out of paper examples of cones like those shown above. What happens to the triangles and lines on a 450° cone? Is the shortest path always straight? Does every pair of points determine a straight line?

Finally, also consider line symmetries on the cone and cylinder. Check to see if the symmetries you found on the plane will work on these surfaces, and remember to think intrinsically and locally. A special class of geodesics on the cone and cylinder are the generators. These are the straight lines that go through the cone point on the cone or go parallel to the axis of the cylinder. These lines have some extrinsic symmetries (can you see which ones?), but in general, geodesics have only local, intrinsic symmetries. Also, on the cone, think about the region near the cone point — what is happening there that makes it different from the rest of the cone?

It is best if you experiment with paper models to find out what geodesics look like on the cone and cylinder before going on to the next page. You will benefit much more from thinking about this problem on your own first, even if you make mistakes.

Geodesics on Cylinders

Let's first look at the three classes of straight lines on a cylinder.

When walking on the surface of a cylinder, a bug might walk along a vertical generator.

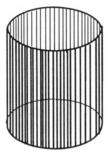

Figure 4.6   **Vertical generators are straight.**

It might walk along an intersection of a horizontal plane with the cylinder, what we will call a *great circle* or a *generator circle*.

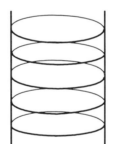

Figure 4.7   **Generating circles are intrinsically straight.**

Or, the bug might walk along a spiral or helix of constant slope around the cylinder.

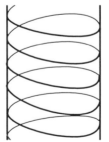

Figure 4.8   **Helixes are intrinsically straight.**

Why are these geodesics? How can you convince yourself? And why are these the only geodesics?

Geodesics on Cones

Now let's look at the classes of straight lines on a cone.

Walking along a generator: When looking at straight paths on a cone, you will be forced to consider straightness at the cone point. You might decide that there is no way the bug can go straight once it reaches the cone point, and thus a straight path leading up to the cone point ends there. Or you might decide that the bug can find a continuing path that has at least some of the symmetries of a straight line. Do you see which path this is?

Bug walking straight over the cone point.   **Figure 4.9**

Walking straight and around: If you use a ribbon on a 90° cone, then you can see that this cone has a geodesic like the one depicted in Figure 4.10. This particular geodesic intersects itself. However, check to see that this property depends on the cone angle. In particular, if the cone angle is more than 180°, then geodesics do not intersect themselves. And if the cone angle is less than 90°, then geodesics (except for generators) intersect at least two times. Try it out! Later in this chapter we will describe a tool which will help you determine how the number of self-intersections depends on the cone angle.

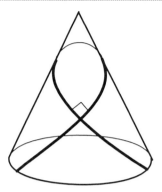

Figure 4.10   **A geodesic intersecting itself on a 90° cone.**

Locally Isometric

By now you should realize that when a piece of paper is rolled or bent into a cylinder or cone, the bug's local and intrinsic experience of the surface does not change except at the cone point. Extrinsically, the piece of paper and the cone are different, but in terms of the local geometry intrinsic to the surface they differ only at the cone point.

Two geometric spaces, G and H, are said to be *locally isometric* at points G in G and H in H if the local intrinsic experience at G is the same as the experience at H. That is, there are neighborhoods of G and H that are identical in terms of intrinsic geometric properties. A cylinder and the plane are locally isometric (at every point) and the plane and a cone are locally isometric except at the cone point. Two cones are locally isometric at their cone points only if the cone angles are the same.

Euclid defines a right angle as follows: *When a straight line set up on a straight line makes the adjacent angles equal to one another, each of the equal angles is* **right** [**A**: Euclid's *Elements*]. Note that if you use this definition, then right angles at a cone point are not equal to right angles at points which are locally isometric to the plane. And Euclid goes on to state as a postulate: *All right angles are equal to one another.* Thus, Euclid's Postulate rules out cone points.

[Problem 5a and the remainder of Chapter 4 are not used in the following chapters.]

PROBLEM *5a. Covering Spaces and Global Properties of Geodesics*

Now we will look more closely at long geodesics that wrap around on a cylinder or cone. Several questions have arisen:

> *How many times can a geodesic intersect itself? How are the self-intersections related to the cone angle? At what angle does the geodesic intersect itself? How can we justify this relationship?*

> *How do we determine the different geodesics connecting two points? How many are there? How does it depend on the cone angle? Is there always at least one geodesic joining each pair of points? How can we justify our conjectures?*

> *What kind of tool could we equip the bug with to help with this investigation?*

Suggestions

Here we offer the tool of covering spaces which may help you explore these questions. The method of "coverings" is so named because it utilizes layers (or sheets) that each "cover" the surface. We will first start with a cylinder because it's easier, and then move on to a cone.

n-Sheeted Coverings of a Cylinder

To understand how the method of coverings works, imagine taking a paper cylinder and cutting it axially (along a vertical generator) so that it unrolls into a plane. This is probably the way you constructed cylinders to study this problem before. The unrolled sheet (a portion of the plane) is said to be a *1-sheeted covering* of the cylinder. See Figure 4.11. If you marked two points on the cylinder, *A* and *B*, as indicated in the figure, then when the cylinder is cut and unrolled into the covering, these two points become two points on the covering (which are labeled by the same letters in the figure). The two points on the covering are said to be *lifts* of the points on the cylinder.

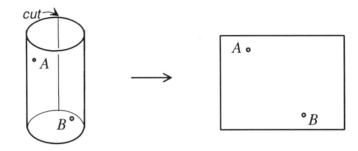

Figure 4.11   **A 1-sheeted covering of a cylinder.**

Now imagine attaching several of these "sheets" together, end-to-end. When rolled up, each sheet will go around the cylinder exactly once — they will each cover the cylinder. (Rolls of toilet paper or paper towels give a rough idea of coverings of a cylinder.) Also, each sheet of the covering will have the points A and B in identical locations. You can see this (assuming the paper thickness is negligible) by rolling up the coverings and making points by sticking a sharp object through the cylinder. This means that all the A's are coverings of the same point on the cylinder and all the B's are coverings of the same point on the cylinder. We just have on the covering several representations, or **lifts**, of each point on the cylinder. Figure 4.12 depicts a 3-sheeted covering space for a cylinder and six geodesics joining A to B. (One of them is the most direct path from A to B and the others spiral once, twice, or three times around the cylinder in one of two directions.)

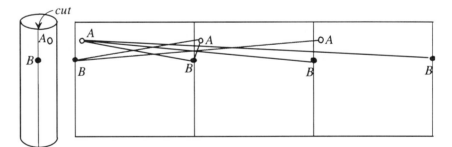

Figure 4.12   **A 3-sheeted covering space for a cylinder.**

We could also have added more sheets to the covering on either the right or left side. You can now roll these sheets back into a cylinder and see what the geodesics look like. Remember to roll it up so that each sheet of the covering completely covers the cylinder — all of the vertical lines between the coverings should lie on the same generator of the cylinder. Note that if you do this with ordinary paper, part or all of some geodesics will be hidden, even though they are all there. It may be easier to see what's happening if you use transparencies.

This method works because straightness is a local intrinsic property. Thus lines that are straight when the coverings are laid out in a plane will still be straight when rolled into a cylinder. Remember that bending the paper does not change the intrinsic nature of the surface. Bending only changes the curvature that we see extrinsically. It is important to always look at the geodesics from the bug's point of view. The cylinder and its covering are locally isometric.

Use coverings to investigate Problem 5a on the cylinder. The global behavior of straight lines may be easier to see on the covering.

n-Sheeted Coverings of a Cone

Figure 4.13 shows a 1-sheeted covering of a cone. The sheet of paper and the cone are locally isomorphic except at the cone point. The cone point is called a ***branch point*** of the covering. We talk about lifts of points on the cone in the same way as on the cylinder. In Figure 4.13 we depict a 1-sheeted covering of a 135° cone and label two points and their lifts.

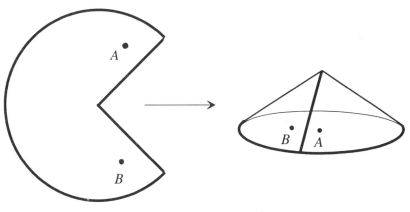

1-sheeted covering of a 135° cone.   **Figure 4.13**

A 4-sheeted covering space for a cone is depicted in Figure 4.14. Each of the rays drawn from the center of the covering is a lift of a single ray on the cone. Similarly, the points marked on the covering are the lifts of the points *A* and *B* on the cone. In the covering there are four segments joining a lift of *A* to different lifts of *B*. Each of these segments is the lift of a different geodesic segment joining *A* to *B*.

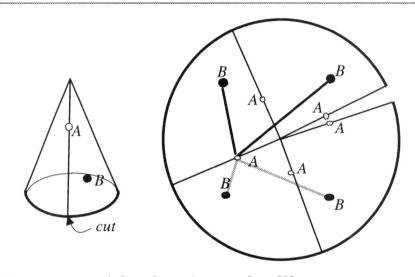

Figure 4.14   **4-sheeted covering space for a 89° cone.**

Think about ways that the bug can use coverings as a tool to expand its exploration of surface geodesics. Also, think about ways you can use coverings to justify your observations in a intrinsic way. It is important to be precise; you don't want the bug to get lost! Count the number of ways in which you can connect two points with a straight line and relate those countings with the cone angle. Does the number of straight paths only depend on the cone angle? Look at the 450° cone and see if it is always possible to connect any two points with a straight line. **Make paper models!** It is not possible to get an equation that relates the cone angle to the number of geodesics joining every pair of points. However, it is possible to find a formula that works for most pairs.

Make covering spaces for cones of different size angles and refine the guesses you have already made about the numbers of self-intersections.

In studying the self-intersections of a geodesic *l* on a cone, it may be helpful for you to consider the ray *R* such that the line *l* is perpendicular to it. (See Figure 4.15.) Now study one lift of the geodesic *l* and its relationship to the lifts of the ray *R*. Note that the seams between individual wedges are lifts of *R*.

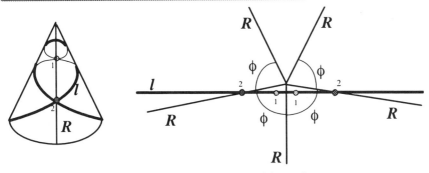

Self-intersections on a cone with angle φ.   Figure 4.15

Pause, experiment, and write out your thoughts.

Covering Space of the Flat Torus

Another example of a covering space is provided by a video game that was popular a while ago. A blip on the video screen representing a ball travels in a straight line until it hits an edge of the screen. Then, the blip reappears traveling parallel to its original direction from a point at the same height on the opposite edge. Is this a representation of some surface? If so, what surface? First, imagine rolling the screen into a tube where the top and bottom edges are joined. This is a representation of the screen as a one-sheeted covering of the cylinder. A blip on the screen that goes off the top edge and reappears on the bottom is the lift of a point on the cylinder which travels around the cylinder crossing the line which corresponds to the joining of the top and bottom of the screen.

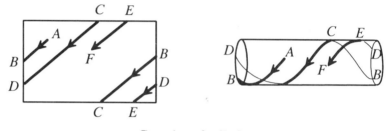

Figure 4.16   Covering of cylinder.

Now, let us further imagine that the cylinder can be stretched and bent so that we can join the two ends to make a torus. Now the screen represents a one-sheeted covering of the torus. If the blip goes off on one side and comes back on the other at the same height, this represents the lift of a point moving around the torus and crossing the circle which corresponds to the place where the two ends of the cylinder are joined. The possible motions of a point on the torus are represented by the motions on the video screen!

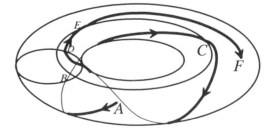

Figure 4.17   Flat torus.

You can't make a model in 3-space of a torus from a flat piece of paper without distorting it, but you can in 4-space! Such a torus is called a *flat torus*. Note that the four corners of the computer screen are lifts of the same point and that a neighborhood of this point has 360° — 90° from each corner.

Coverings and the Sphere

There is no way to construct a covering of a sphere which has more than one sheet unless the covering has some "branch points." A *branch point* on a covering is a point such that every neighborhood (no matter how small) surrounding the point contains at least two lifts of some point. In any covering of a cone with more than one sheet, the lift of the cone point is a branch point as you can see in Figure 4.18.

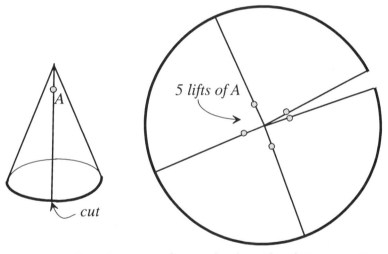

Covering space of a cone has branch points.   **Figure 4.18**

Notice that the coverings of a cylinder and a torus have no branch points. For a sphere the matter is very different — any covering of a sphere will have a branch point. You can see this if you try to construct a cover by slitting the sphere as depicted in Figure 4.19. The ends of this slit would become branch points. This topic may be explored further in geometric or algebraic topology.

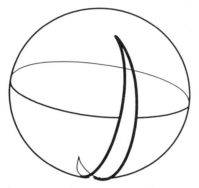

Figure 4.19   **Covering space of a sphere has branch points.**

In fact, any surface which has no (non-branched) coverings and which is bounded and without an edge can be continuously deformed (without tearing) into a round sphere. The surfaces of closed boxes and of footballs are two examples. A torus is bounded and without an edge, but it cannot be deformed into a sphere. A cylinder also cannot be deformed into a sphere, and a cylinder either has an edge or (if we imagine it as extending indefinitely) it is unbounded.

A 3-dimensional analog of this situation arises from a famous, long-unsolved problem called the *Poincaré Conjecture*. The analog of a surface is called a *3-dimensional manifold*, a space which is locally like Euclidean 3-space, (in the same sense that a surface is locally like the plane). The 3-sphere, which we will study in Chapter 11, is a 3-dimensional manifold. Also, our physical 3-dimensional universe may not be Euclidean, but, certainly, it is locally like Euclidean 3-space, and thus is a 3-dimensional manifold. Poincaré (a famous mathematician in the early part of this century) conjectured that any 3-dimensional manifold which has no (non-branched) coverings and which is bounded and without boundary must be deformable into a 3-dimensional sphere. For the past 80 years, numerous mathematicians have tried to decide whether Poincaré's conjecture is true or not. So far, no one has succeeded. See [O: Hilbert] and [C: Weeks] for more discussion of 3-dimensional manifolds and the 3-dimensional sphere.

Exploring covering spaces brings out differences among the plane, sphere, cylinder, and cone in a very natural way. You should realize several different features of each surface. For example, a cylinder has (un-branched) coverings and is locally isometric to the plane. A cone has only branched coverings and is locally isometric to the plane except at the cone point. Like the cone the sphere also has only branched coverings but we will see later that the sphere is not locally isometric to the plane.

Is "Shortest" Always "Straight"?

We are often told that "a straight line is the shortest distance between two points," but, is this really true?

As we have already seen on a sphere, two points which are not opposite each other are connected by two straight paths (one going one way around a great circle and one going the other way). Only one of these paths is shortest. The other is also straight, but not the shortest straight path.

Consider a model of a cone with angle 450°. Notice that such cones appear commonly in buildings as so-called "outside corners" (see Figure 4.20). It is best, however, for you to have a paper model that can be flattened. Use your model to investigate which points on the cone can be joined by straight lines. In particular, look at points like those labeled *A* and *B* in Figure 4.20 below. There is no single straight line on the cone going from *A* to *B*, and thus for these points the shortest path is not straight. Convince yourself that in this case this shortest path is not straight.

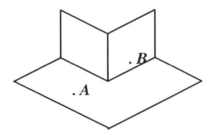

There is no straight path from *A* to *B*.   Figure 4.20

Here is another example: Think of a bug crawling on a plane with a tall box sitting on that plane (refer to Figure 4.21). This combination surface — the plane with the box sticking out of it — has eight cone points. The four at the top of the box have 270° cone angles, and the four at the bottom of the box have 450° cone angles (180° on the box and 270° on the plane). What is the shortest path from points *X* and *Y*, points which are on opposite sides of the box? Is the straight path the shortest? Is the shortest path straight? To check that the shortest path is not straight, try to see that at the bottom corners of the box the two sides of the path have different angular measures. (In particular, if *X* and *Y* are close to the box, then the angle on the box side of the path measures a little more than 180° and the angle on the other side measures almost 270°.)

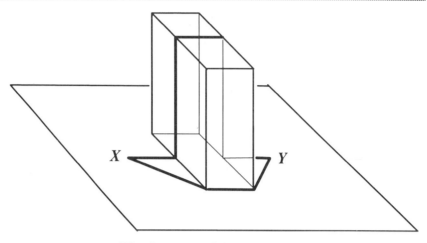

Figure 4.21   **The shortest path is not straight.**

So, we see that sometimes a straight path is not shortest and the shortest path is not straight. However, for "smooth" surfaces, "straight" and "shortest" are more nearly the same. A *smooth* surface is essentially what it sounds like. More precisely, a surface is smooth at a point if, when you zoom in on the point, the surface becomes indistinguishable from a flat plane.[†] Note that a cone is not smooth at the cone point, but a sphere and a cylinder are both smooth at every point. The following is a theorem from differential geometry:

> **THEOREM:** *If a surface is smooth then a straight line on the surface is always the shortest path between "nearby" points. If the surface is smooth and complete (every geodesic on it can be extended indefinitely), then the shortest path between any two points is always straight.*

Consider a planar surface with a hole removed. Check that for points near opposite sides of the hole, the shortest path (on the plane surface with hole removed) is not straight because the shortest path must go around the hole.

[†] In particular, let p denote a point on the surface and let $S(\varepsilon)$ be that portion of the surface which is inside a small sphere whose center is p and whose radius is ε. Now, magnify $S(\varepsilon)$ by $1/\varepsilon$ so that it now has radius 1; denote this magnified piece of the surface $1/\varepsilon \times S(\varepsilon)$. Then the surface is smooth at p if $1/\varepsilon \times S(\varepsilon)$ becomes indistinguishable from a portion of a (flat) plane when ε goes to zero.

Chapter 5
SAS and ASA

Let the following be postulated: to draw a straight line from any point to any point.

— Euclid, *Elements*, Postulate 1 [**A**: Euclid]

At this point, you should be thinking intrinsically about the surfaces of spheres, cylinders, and cones. In the problems to come you will have opportunities to apply your intrinsic thinking when you make your own definitions for triangle on the different surfaces and investigate congruence properties of triangles.

Each of the two congruence properties that you will investigate in Problems 6 and 7, "Side-Angle-Side" and "Angle-Side-Angle," can be seen as an expression of Euclid's first postulate quoted above. There are two crucial properties that make SAS and ASA true on the plane: (1) *There is only one straight line segment joining two points*, and, (2) *two different straight lines intersect in at most one point*. It has been claimed that these two different expressions follow from the Euclid's Postulate 1. A contemporary reading of Greek does not make it clear whether Euclid actually asserts in his first postulate that the straight line is unique. However, he uses the postulate in his proof of *Side-Angle-Side* in a manner that makes it clear that he considers the line to be unique.

We will now look at triangles on a plane, sphere, cone and cylinder. [If you skipped Chapter 4, you should still find that this and the succeeding chapters will still make sense but you may want to limit your investigations to triangles on planes and spheres.]

In order that we may start out with some common ground, let us agree on some common terminology: A *triangle* has three points (*vertices*) which are joined by three straight line (geodesic) segments (*sides*). A triangle divides the surface into two regions (the *interior* and *exterior*). The (*interior*) *angles* of the triangle are the angles between the sides in the interior of the triangle.

PROBLEM 6. *Side-Angle-Side (SAS)*

> *Are two triangles congruent if two sides and the included angle of one are congruent to two sides and the included angle of the other?*

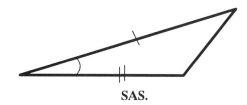

Figure 5.1   **SAS.**

> *In some textbooks SAS is listed as an axiom; in others it is listed as the definition of congruency of triangles, and in others as a theorem to be proved. But no matter how one considers SAS, it still makes sense and is important to ask: Why is SAS true on the plane? One can also ask: Is SAS true on spheres, cylinders, and cones?*

> *If you find that SAS is not true for all triangles on a sphere or another surface, is it true for sufficiently small triangles? Come up with a definition for "small triangles" for which SAS does hold.*

Suggestions

Be as precise as possible, but **use your intuition**. In trying to prove SAS on non-planar surfaces you will realize that SAS does not hold unless some restrictions are made on the triangles. Keep in mind that everyone sees things differently, so there are many possible definitions of "small." Some may be more restrictive than others (i.e., they don't allow as many triangles as other definitions). Use whatever definition makes sense for you.

Remember that it is not enough to simply state what a small triangle is; you must also prove that SAS is true for the small triangles under your definition — explain why the counterexamples you found before are now ruled out and explain why the condition(s) you list is (are) sufficient to prove SAS. Also, try to come up with a basic, general proof that can be applied to all surfaces.

And remember what we said before: By "proof" we mean what most mathematicians use in their everyday practice, i.e., an argument that is sufficient to convince a reasonable skeptic, not the usual two-column proofs from high school (unless, of course, you find the two-column proof sufficiently convincing). Your proof should convey the meaning you are experiencing in the situation. Think about why SAS is true on the plane — think about what it means for actual physical triangles — then try to translate these ideas to a sphere, cone, and cylinder.

Let us clarify some terminology that we have found to be helpful for discussing SAS and other theorems. Two triangles are said to be *congruent* if through a combination of translations, rotations, and reflections, one of them can be made to coincide with the other. If no reflections are needed, then the triangles are said to be *directly congruent*. In this course we will focus on *congruence* and not specifically on *direct congruence*; however, some students may wish to keep track of the distinction as we go along.

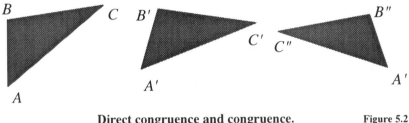

Direct congruence and congruence.   Figure 5.2

In Figure 5.2, $\triangle ABC$ is directly congruent to $\triangle A'B'C'$ but $\triangle ABC$ is not directly congruent to $\triangle A''B''C''$. However, $\triangle ABC$ is congruent to both $\triangle A'B'C'$ and $\triangle A''B''C''$ and we write: $\triangle ABC \cong \triangle A'B'C' \cong \triangle A''B''C''$.

So, why is SAS true on the plane? We will now illustrate one way of looking at this question. Referring to Figure 5.3, suppose that $\triangle ABC$ and $\triangle A'B'C'$ are two triangles such that $\angle BAC \cong \angle B'A'C'$, $AB \cong A'B'$ and $AC \cong A'C'$. Translate $\triangle A'B'C'$ along AA' so that A' coincides with A. Since the sides AC and $A'C'$ are congruent we can now rotate $\triangle A'B'C'$ (about $A=A'$) until C' coincides with C. If after this rotation B and B' are not coincident, then a reflection (about $AC = A'C'$) will complete the process and all three vertices, the two given sides, and the included angle of the two triangles will coincide.

So, why is it that, on the plane, the third sides (BC and B'C') must now be the same? Since the third sides (BC and $B'C'$) coincide, $\triangle ABC$ is congruent to $\triangle A'B'C'$. (In the case that no reflection is needed, the two triangles are directly congruent.)

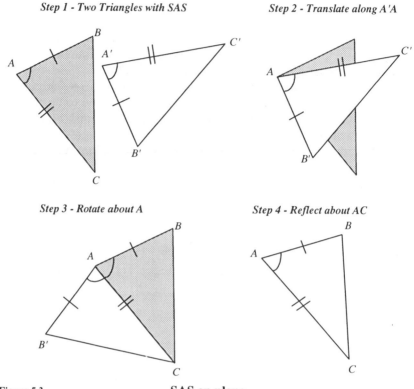

Step 1 - Two Triangles with SAS

Step 2 - Translate along A'A

Step 3 - Rotate about A

Step 4 - Reflect about AC

Figure 5.3   **SAS on plane.**

The proof of SAS on the plane is not directly applicable to the other surfaces because geodesics differ from surface to surface. In particular, the number of geodesics joining two points varies from surface to surface and is also relative to the location of the points on the surface. On a sphere, for example, there are always at least two straight paths joining any two points. As we saw in Chapter 4, the number of geodesics joining two points on a cylinder is infinite. On a cone the number is dependent on the cone angle, but for cones with angles less than 180° there is more than one geodesic joining two points. It follows that the argument made for SAS on the plane is not valid on cylinders, cones, or spheres. The question then arises: Is SAS *ever* true on those surfaces?

Look for triangles for which SAS is not true. Some of the properties that you found for geodesics on spheres, cones, and cylinders will come into play. As you look closely at the features of triangles on those surfaces, you may find that they challenge your notions of triangle. Your intuitive notion of triangle may go beyond what can be put into a traditional definition of triangle. When you look for a definition of *small triangle* for which SAS will hold on these surfaces, you should try to stay close to your intuitive notion. In the process of exploring different triangles you may come up with examples of triangles that seem very strange. Let's look at some strange triangles.

For instance, keep this example in mind:

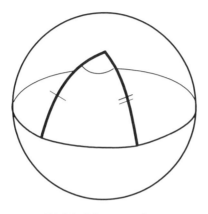

SAS is false on sphere.   **Figure 5.4**

All the lines shown in Figure 5.4 are geodesic segments of the sphere. The two sides and the included angle for SAS are marked. As you can see, there are two possible geodesics that can be drawn for the third side—the short one in front and the long one that goes around the back of the sphere. Remember that on a sphere, any two points define at least two geodesics (an infinite number if the points are at opposite poles). Look for similar examples on a cone and cylinder. You may decide to accept the smaller triangle into your definition of "small triangle" but to exclude the large triangle from your definition. But what is a large triangle? To answer this, let us go back to the plane. What is a triangle on the plane? What do we choose as a triangle on the plane?

On the plane what we want to call a triangle has all of its angles on the "inside." Also, there is a clear choice for *inside* on the plane; it is the side that has finite area. See Figure 5.5. But what is the inside of a triangle on a sphere? The restriction that the area on the inside has to be finite doesn't work for the spherical triangles because all areas on a sphere are finite. So what is it about the large triangle that challenges our view of triangle? You might try to resolve the triangle definition problem by specifying that each side must be the shortest geodesic between the endpoints. However, be aware that antipodal points (that is, a pair of points that are at diametrically opposite poles) on a sphere do not have a unique shortest geodesic joining them. On a cylinder we can have a triangle for which all the sides are the shortest possible segments, yet the triangle does not have finite area. Try to find such an example. In addition, a triangle on a cone will always bound one region that has finite area. Look at some of these ornery examples of triangles. A triangle that encircles the conepoint may cause problems. Covering spaces can help you in your investigation of these triangles. For example, what happens when we try to unwrap or lift one of these triangles onto a covering space?

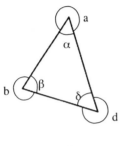

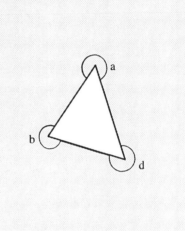

Figure 5.5   **Insides of a plane triangle.**

PROBLEM 7. Angle-Side-Angle (ASA)

Are two triangles congruent if one side and the adjacent angles of one are congruent to one side and the adjacent angles of another?

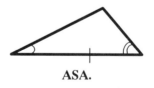

Figure 5.6   **ASA.**

Suggestions

This problem is similar in many ways to the previous one. As before, look for counterexamples on all surfaces, and if ASA doesn't hold for all triangles, see if it works for small triangles. If you find that you must restrict yourself to small triangles, see if your previous definition of "small" still works; if it doesn't work here, then modify it.

There are also a few things to keep in mind while working on this problem. First, when considering ASA, both of the angles must be on the same side — the interior of the triangle. For example,

This is not an example of ASA.   But this is an example of ASA.

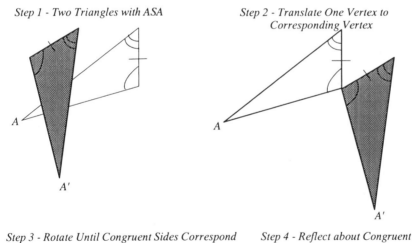

Angles of a triangle must be on same side.   Figure 5.7

Let us look at a proof of ASA on the plane:

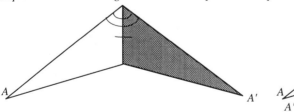

ASA on the plane.   Figure 5.8

The planar argument for ASA does not work on spheres, cylinders, and cones because, in general, geodesics on these surfaces intersect in more than one point. As was the case for SAS, we must ask ourselves if we can find a class of small triangles on each of the different surfaces for which the above argument is valid. You should check if your previous definitions of small triangle are too weak, too strong, or just right to make ASA true on spheres, cylinders, and cones. It is also important to look at cases for which ASA does not hold. Just as with SAS, some interesting counterexamples arise.

In particular, try out this configuration on a sphere. To see what happens you will need to try this on an actual sphere.

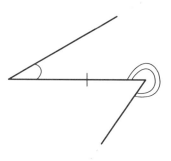

Figure 5.9   **Possible counterexample to ASA.**

If you extend the two sides to great circles, what happens? You may instinctively say that it is not possible for this to be a triangle, and on the plane most people would agree, but *try it on a sphere and see what happens*. Does it define a unique triangle? Remember that on a sphere two geodesics always intersect twice.

Finally, notice that in our proof of ASA on the plane, we did not use the fact that the sum of the angles in a triangle is 180°. We avoided this for two reasons. For one thing, to use this "fact" we would have to prove it first. This is both time consuming and unnecessary. We will prove it later (Problem 23). The other reason is that a proof using the fact that the angles sum to 180° will not work on a sphere because there are at least some triangles on a sphere whose angles sum to more than 180°. A common example is the triple-right triangle, depicted in Figure 5.10, which you may have seen before.

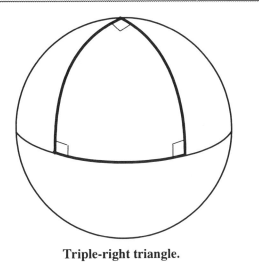

Triple-right triangle.   Figure 5.10

Remember that it is best to come up with a proof that will work for all surfaces because this will be more powerful, and, in general, will tell us more about the relationship between the plane and the other surfaces.

Pause, explore, and write out your ideas for these problems before going to the next page.

Addendum on the Use of Covering Spaces

It is natural to be uncomfortable using covering spaces, but covering spaces are a helpful tool for thinking intrinsically. Some triangles, even though they look strange extrinsically, will look like reasonable triangles for the bug. Below we give an example of an extrinsically strange triangle which intersects itself but which can be considered a normal triangle from an intrinsic point of view. In fact, it is a planar triangle. Such triangles have all the properties of plane triangles including SAS and ASA.

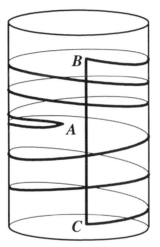

Unroll to a 3-sheeted cover, and ...

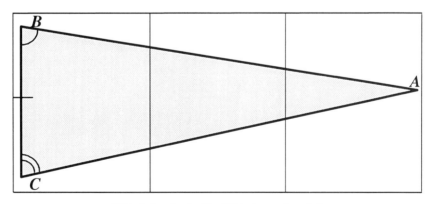

Figure 5.11   **Think intrinsically. This is a triangle!**

Chapter 6
Area, Parallel Transport, and Holonomy

> **Parallel** straight lines are straight lines which, being in the same place and being produced indefinitely in both directions, do not meet one another in either direction.
>
> — Euclid, *Elements*, Definition 23 [**A**: Euclid]

In this chapter we will find a formula for the area of a triangle on a sphere. We will then investigate the connections between area and *parallel transport*, a notion of local parallelism that is definable on all surfaces. We will also introduce the notion of *holonomy* which has many applications in modern differential geometry and engineering.

PROBLEM 8. *The Area of a Triangle on a Sphere*

Let Δ *denote a triangle on a sphere. Show that the formula*

$$\text{Area} (\Delta) = [\ \sum \angle\text{'s} - \pi\ ]\ A/4\pi$$

makes sense, where A *is the area of the sphere and* $\sum \angle\text{'s}$ *is the sum of the angles of* Δ *in radians.*

Suggestions

The quantity $(\sum \angle\text{'s}) - \pi$ is called the *excess* of Δ. We offer the following hint as a way to approach this problem: Find the area of a biangle (lune) with angle α. (A *biangle* or *lune* is the region between two great circles.) Notice that the great circles which contain the sides of the triangle divide the sphere into overlapping biangles (in more than one way).

This is one of the problems that you almost certainly must do on an actual sphere. There are simply too many things to see, and the drawings we make on paper distort lines and angles too much. The best way to start is to make a small triangle on a sphere, and extend the sides of the triangle to complete great circles. Then look at what you've got. You will find an identical triangle on the other side of the sphere, and you can see several lunes that extend out from the triangles. The key to this problem is to put everything in terms of areas that you know. This is where the hint comes in — find the areas of the lunes. After that, it is simply a matter of adding everything up properly.

We will see later (Problem 43) that the area of the whole sphere is $4\pi R^2$, where R is the (extrinsic) radius of the sphere. With this additional information we can rewrite the formula of Problem 8:

$$\text{Area } (\Delta) = [\ \sum \angle\text{'s} - \pi\ ]\ R^2.$$

Introducing Parallel Transport and Holonomy

Imagine that you are walking along a straight line or geodesic carrying a stick that makes a fixed angle with the line you are walking on. If you walk along the line maintaining the direction of the stick relative to the line constant, then you are performing a *parallel transport* of that "direction" along the path.

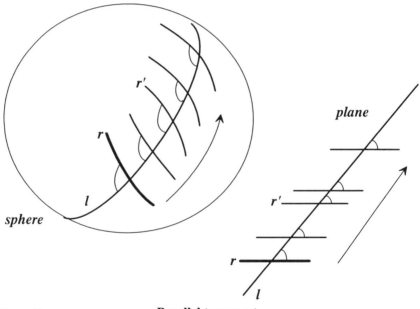

Figure 6.1   **Parallel transport.**

To express the parallel transport idea, it is common terminology to say that:

♦ r' is a parallel transport of r along l;

♦ r is a parallel transport of r' along l;

◆   *r* and *r'* are parallel transports along *l*;

◆   *r* can be parallel transported along *l* to *r'*; or,

◆   *r'* can be parallel transported along *l* to *r*.

On the plane there is a global notion of parallelism — two lines in the same plane are ***parallel*** if they do not intersect when extended. It follows from Problem 16 (or standard results in high school geometry) that if two lines are parallel transports along another line in the plane, then they are parallel in the sense that they will not intersect if extended. On a sphere this is not true — any two great circles on the same sphere intersect, and intersect twice. Also, in Problem 20 you will show that if two lines on the plane are parallel transports along a third line, then they are parallel transports along every line that transverses them. This is also not true on a sphere. For example, any two great circles (longitudes) through the north pole are parallel transports of each other along the equator, but they are not parallel transports along great circles near the North Pole. We will explore this aspect of parallel transport more in Chapters 8 and 9.

Parallel transport has become an important notion in differential geometry, physics, and mechanics. One important aspect of differential geometry is the study of properties of spaces (surfaces) from an intrinsic point of view. As we have seen, in general, it is not possible to have a global notion of direction from which we are able to determine when a direction (or vector) at one point is the same as a direction (or vector) at another point. However, we can say that they have the same direction *with respect to* a geodesic *g* if they are parallel transports of each other along *g*. Parallel transport can be extended to arbitrary curves as we shall discuss after Problem 11. With this notion it is possible to talk about how a particular vector quantity changes intrinsically along a curve (covariant differentiation). In general, covariant differentiation is useful in the areas of physics and mechanics. In physics, the notion of parallel transport is central to some of the theories that have been put forward as possible candidates for a "Unified Field Theory," a hoped for but as yet unrealized theory that would unify all known physical laws about forces of nature.

Now let us explore what happens when we parallel transport a line segment around a triangle. For example, consider on a sphere an isosceles triangle with base on the equator and opposite vertex on the North Pole. See Figure 6.2. Note that the base angles are right angles. Now start at the North Pole with a vector (a directed geodesic segment) and parallel transport it along one of the sides of the triangle until it reaches the base. Then parallel transport it along the base to the third side. Then parallel transport back to the North Pole along the third side. Notice that the vector now points in a different direction then it did originally. This difference is called the ***holonomy*** of the triangle.

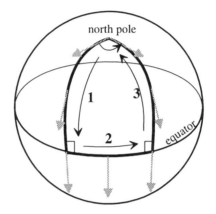

Figure 6.2   **The holonomy of a double-right triangle.**

This works on the sphere for any small triangle (i.e., a triangle that is contained in an open hemisphere). We can define **the holonomy of a small triangle**, $\mathscr{H}(\Delta)$, as follows:

> *If you parallel transport a vector (a directed geodesic segment) counterclockwise around the three sides of a small triangle, then the holonomy of the triangle is the smallest angle measured counterclockwise from the original position of the vector to its final position.*

Holonomy can also be defined for large triangles but it is more complicated because of the confusion as to what angle to measure. For example, what should be the holonomy when you parallel transport around the equator — zero radians or 2π radians?

PROBLEM 9. *The Holonomy of a Small Triangle*

> *Find a formula that expresses the holonomy of a small triangle on a sphere.*

Suggestions

What happens to the holonomy when you change the angle at the North Pole of the triangle in Figure 6.2? What happens if you parallel transport around the triangle a vector pointing in a different direction? Parallel transport vectors around different triangles on your model of a sphere. Try it on triangles that are very nearly the whole hemisphere and try it on very small triangles. What do you notice?

A good way to approach the formula for general small triangles is to start with any geodesic segment at one of the angles of the triangle, and follow it as it is parallel transported around the triangle. Keep track of the relationships between the angles this segment makes with the sides and the exterior angles. See Figure 6.3

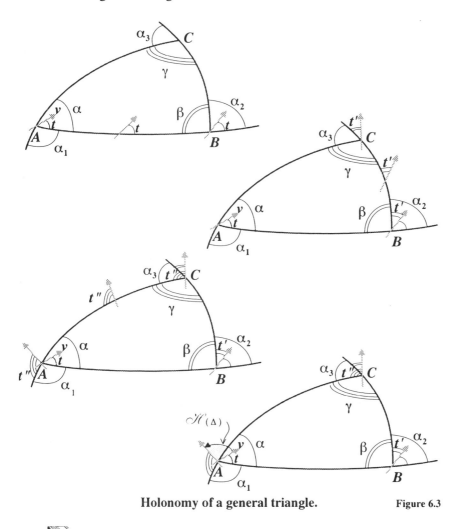

Holonomy of a general triangle.   Figure 6.3

Pause, explore, and write out your ideas for this problem before going on to the next page.

The Gauss-Bonnet Formula for Small Triangles

In working on Problem 9 you should find (among other things) that:

> *The holonomy of a small triangle is equal to 2π minus the sum of the exterior angles and is the same as its excess.*

Let α, β, and γ be the interior angles of the triangle and α_1, α_2, α_3 the exterior angles. Then algebraically the statement above can be written as:

$$\mathcal{H}(\Delta) = 2\pi - (\alpha_1 + \alpha_2 + \alpha_3) = (\alpha + \beta + \gamma) - \pi$$

Note that one consequence of this formula is that the holonomy does not depend on either the vertex or the vector we start with. This is to be expected since parallel transport does not change the relative angles of any figure.

Following Problem 8 we can write the result in this form:

$$\mathcal{H}(\Delta) = 2\pi - (\alpha_1 + \alpha_2 + \alpha_3) = A\ (\Delta)\ 4\pi/A = A\ (\Delta)\ R^{-2}$$

The formula $2\pi - (\alpha_1 + \alpha_2 + \alpha_3) = A\ (\Delta)\ R^{-2}$ is called the **Gauss-Bonnet Formula** (for triangles). The quantity R^{-2} is traditionally called the **Gaussian curvature** or just plain **curvature** of the surface.

Can you see how this result gives the bug an intrinsic way of determining the extrinsic quantity R and the curvature R^{-2}?

The Gauss-Bonnet Formula not only holds for small triangles but can also be extended to any small (that is, contained in an open hemisphere) simple (i.e., non-intersecting) polygon (that is, a closed curve made up of a finite number of geodesic segments) contained on a sphere.

The holonomy of a small simple polygon, $\mathcal{H}(\Gamma)$, is defined as follows:

> *If you parallel transport a vector (a directed geodesic segment) counterclockwise around the sides of a small simple polygon, then the holonomy of the polygon is the smallest angle measured counterclockwise from the original position of the vector and its final position.*

[Problems 10 and 11 are not necessary for the remainder of the book and therefore the rest of this chapter may be skipped now if you wish.]

PROBLEM 10. *The Gauss-Bonnet Formula for Polygons on a Sphere*

If you walk around a polygon with the interior of the polygon on the left, the exterior angle at a vertex is the change in the direction at that vertex. This change is positive if you turn counterclockwise and negative if you turn clockwise. (See Figure 6.4.)

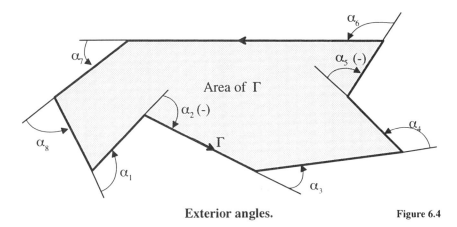

Exterior angles.   Figure 6.4

Show that if Γ *is a small simple polygon on a sphere, then*

$$\mathcal{H}(\Gamma) = 2\pi - \Sigma\, \alpha_i = A\ (\Gamma)\ 4\pi/A = A\ (\Gamma)\ R^{-2},$$

where $\Sigma\alpha_i$ *is the sum of the exterior angles of the polygon.*

Suggestions

Outline of a proof: Divide the polygon into small triangles. It is possible to do this by constructing geodesic segments in the interior of the polygon without adding any new vertices (see Problem 11). For now, assume (do not prove) that this is possible. Do this problem in two steps: First, by removing the small triangles one at a time, show that the holonomy of the polygon is the sum of the holonomies of the triangles. Then, second, check directly that $\mathcal{H}(\Gamma) = 2\pi - \Sigma\, \alpha_i$.

PROBLEM 11. *Dissection of Polygons into Triangles*

Prove that every simple polygon can be dissected into small triangles without adding extra vertices.

Suggestions

Look at this on both the plane and sphere. The difficulty in this problem is coming up with a method that works for all polygons, including very general or complex ones, like this:

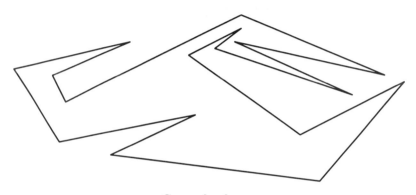

Figure 6.5   **General polygon.**

You may be tempted to try to connect nearby vertices to create triangles, but how do we know that this is always possible? How do you know that in any polygon there is even one pair of vertices that can be joined in the interior? The polygon may be so complex that parts of it get in the way of what you're trying to connect. So you might start by giving a convincing argument that there is at least one pair of vertices that can be joined by a segment in the interior of the polygon. In order to see that there is something to prove here, Figure 6.6 shows an example of a polyhedron in 3-space which has **no** pair of vertices that can be joined in the interior. The polyhedron consists of eight triangular faces and six vertices. Each vertex is joined by an edge to four of the other vertices and the straight line segment joining it to the fifth vertex lies in the exterior of the polygon. Therefore it is impossible to dissect this polyhedron into tetrahedra without adding extra vertices. This example and some history of the problem are discussed in [**P**: Eves, page 211].

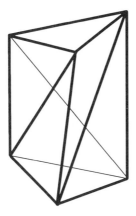

A polyhedron with vertices not joinable in the interior.   Figure 6.6

Gauss-Bonnet Formula on Surfaces

The above discussion of holonomy is in the context of a sphere, but the results have a much more general applicability and constitute a major aspect of differential geometry. In particular, we can extend this result even further to general surfaces, even those of non-constant curvature. In fact, the *(Gaussian) curvature* $K(p)$ at a point p on any surface can be defined as,

$$K(p) = \lim_{\Delta \to 0} \mathscr{H}(\Delta) \,/\, A(\Delta),$$

where the limit is taken over all small (geodesic) triangles that contain p. It can be shown using differential geometry that for a surface embedded in 3-space, the sphere that best fits the surface at p is the sphere with radius $(K(p))^{-1/2}$, if $K(p)$ is positive. If a surface at a point has negative curvature (and thus negative holonomy for small triangles containing the point), then near the point the surface looks like a saddle surface. This definition leads us to another formula, namely,

The Gauss-Bonnet Formula for Polygons on Surfaces

On any smooth surface (2-manifold), if Γ is a (geodesic) polygon that is contractible to a point in its interior, then

$$\mathscr{H}(\Gamma) = 2\pi - \Sigma\, \alpha_i \; = \iint_{I(\Gamma)} K \, dA,$$

where the integral is the (surface) integral over $I(\Gamma)$, the interior of the polygon.

The proof of this formula involves dividing the interior of Γ into many triangles, each so small that the curvature K is essentially constant over its interior, and then applying the Gauss-Bonnet Formula for spheres to each of the triangles.

In addition, all of the versions of the Gauss-Bonnet Formula given thus far can be extended to arbitrary, simple, piecewise smooth, closed curves. If γ is such a curve, then we can define the holonomy $\mathcal{H}(\gamma) = \lim \mathcal{H}(\gamma_i)$, where the limit is over a sequence (which converges point-wise to γ) of geodesic polygons $\{\gamma_i\}$ whose vertices lie on γ. Using this definition, the Gauss-Bonnet formula can be extended even further:

The Gauss-Bonnet Formula for Curves Which Bound a Contractible Region

$$\mathcal{H}(\gamma) \ = \ A(\gamma)\,R^{-2} \text{ (on a sphere)}$$

or,

$$\mathcal{H}(\gamma) \ = \ \iint_{I(\gamma)} K\,dA \text{ (on surfaces)},$$

where $I(\gamma)$ is the interior of the region bounded by γ.

Chapter 7

ITT, SSS, and ASS

... the Power of the World always works in circles, and everything tries to be round.
— Black Elk in *Black Elk Speaks*[†]

Let the following be postulated: to produce a circle with any centre and distance.
— Euclid, *Elements*, Postulate 3 [**A**: Euclid]

The focus of this chapter is the Isosceles Triangle Theorem, or ITT. This result is closely related to circles since the two congruent sides may be considered radii of a circle.

PROBLEM 12. *Isosceles Triangle Theorem (ITT)*

Given a triangle with two of its sides congruent, then are the two angles opposite those sides also congruent?

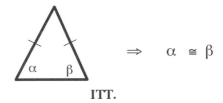

$$\Rightarrow \quad \alpha \cong \beta$$

ITT.   **Figure 7.1**

It may be helpful to use SAS and/or ASA to prove ITT. Looking for symmetry is also very helpful, and it can be used in conjunction with or instead of the previous triangle congruence theorems. Look for counter-examples on a sphere and cylinder, and specifically look at the examples given for ASA. As before, if ITT doesn't work for all triangles, see if it will work for small triangles. Note that you may have to change your definition of "small" for use with ITT and the following problems. Finally, don't create a definition for "small triangle" if it's not necessary. This goes for ITT as well as for the later triangle theorems — some of the

[†] *Black Elk Speaks: Being the Life Story of a Holy Man of the Oglala Sioux* as told through John G. Neihardt, University of Nebraska Press, 1961.

theorems may work for all triangles on a particular surface, so don't restrict yourself if it's not necessary. If you think that a particular theorem is not true for large triangles, describe a counterexample.

In your proof of ITT, try to see that you have also proved the following:

> **COROLLARY.** *The bisector of the top angle of an isosceles triangle is also the perpendicular bisector of the base of that triangle.*

You may also want to prove the converse of ITT, which follows:

> **Converse of ITT.** *For all triangles on the plane and spheres, if two angles are congruent, then the sides opposite the angles are also congruent.*

For the proof, use ASA or symmetries.

Circles

The Isosceles Triangle Theorem will be very useful in Problems 13 and 14. It is also used to prove theorems about circles. We define a circle intrinsically.

> *A circle is a geometric figure formed by rotating one endpoint of a geodesic segment about its other endpoint.*

The intrinsic radius of the circle is the rotating segment and the intrinsic center of the circle is the endpoint about which the rotation is fixed. Note that on a sphere every circle has two (intrinsic) centers which are antipodal (and, in general, two different radii).

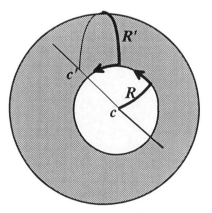

Figure 7.2   **Circles have two centers.**

Now ITT can be used to prove theorems about circles. For example,

> *Two circles on a sphere or a plane intersect in either 0, 1, or 2 points or, if the centers are antipodal and the radii add up to half a great circle, then they coincide.*

Proof: If *a* and *b* are two points of intersection and *c* and *c'* are the centers of the circles, then Δ*acb* and Δ*ac'b* are isosceles triangles. If *c* and *c'* are antipodal then either the circles coincide or are disjoint. Thus, we can assume here that *c* and *c'* are not antipodal, and so there is a unique great circle through the centers. But, given that Δ*acb* and Δ*ac'b* are isosceles, the corollary to ITT asserts that the bisectors of ∠*acb* and ∠*ac'b* must be perpendicular bisectors of their common base. Thus, the union of the two angle bisectors is straight and joins *c* and *c'*. So, the union must be contained in the unique great circle determined by *c* and *c'*. Therefore, any pair of intersections of the two circles, such as *a* and *b*, must lie on opposite sides of this unique great circle. Immediately, it follows that there cannot be more than two intersections unless the circles coincide.

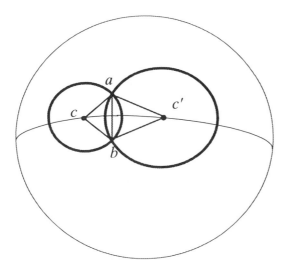

Intersection of two circles.   Figure 7.3

Triangles on Cone and Cylinder

In Problems 6 and 7 you probably noticed that however you defined small triangles for cones and cylinders, they were very much the same as planar triangles. In fact, using coverings, we can say that *any triangle that can be lifted to a covering of a cone or cylinder is a planar triangle* because it is in a covering which is part of the plane. Therefore, any triangle that can be lifted onto a covering must satisfy SAS, ASA, ITT, and any other property of plane triangles. Whatever your favorite definition of small triangles is on cones and cylinders, you should be able to prove, (possibly, with some difficulty) that your small triangles can be lifted to a covering. Therefore, from now on, we will need no special argument for liftable triangles on cones and cylinders — these triangles are plane triangles.

PROBLEM *13. Side-Side-Side (SSS)*

> *Are two triangles congruent if the two triangles have congruent corresponding sides?*

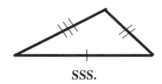

Figure 7.4   SSS.

Suggestions

Start investigating SSS by making two triangles coincide as much as possible, and see what happens. For example, in figure 7.5, if we line up one pair of corresponding sides of the triangles, we have two different orientations for the other pairs of sides:

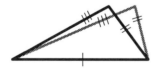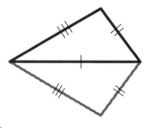

Figure 7.5   **Are these possible?**

Of course, it is up to you to determine if each of these orientations is actually possible, and to prove or disprove SSS. Again, symmetry can be very useful here.

On a sphere, SSS doesn't work for all triangles. The counterexample in Figure 7.6 shows that no matter how small the sides of the triangle are, SSS does not hold because the three sides always determine two different triangles on a sphere. Thus, it is necessary to restrict the size of more than just the sides in order for SSS to hold on a sphere. Whatever argument you used for the plane should work on the sphere for *suitably defined* small triangles. Make sure you see what it is in your argument that doesn't work for large triangles.

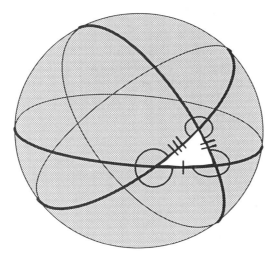

A large triangle with small sides.   Figure 7.6

There are also other types of counterexamples to SSS on a sphere. Can you find them?

PROBLEM 14. Angle-Side-Side (ASS)

Are two triangles congruent if an angle, an adjacent side, and the opposite side of one triangle are congruent to an angle, an adjacent side, and the opposite side of the other?

Figure 7.7   **ASS.**

Suggestions

Suppose you have two triangles with the above congruencies. We will call them ASS triangles. We would like to see if, in fact, the triangles are congruent. We can line up the angle and the first side, and we know the length of the second side (*bc* or *b'c'*), but we don't know where the second and third sides will meet. See Figure 7.8.

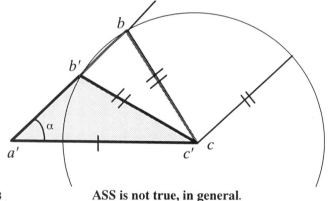

Figure 7.8   **ASS is not true, in general.**

Here, the circle that has as its radius the second side of the triangle intersects the ray that goes from *a* along the angle α to *b* twice. So ASS doesn't work for all triangles on either the plane or a sphere. Try this for yourself on the plane and a sphere to see what happens. Can you make ASS work for an appropriately restricted class of triangles? On a sphere, also look at triangles with multiple right angles, and, again, define "small" triangles as necessary. Your definition of small triangle here may be very different from your definitions in Problems 6 and 7.

There are numerous collections of triangles for which ASS is true. Look at right triangles. Explore. See what you find.

Pause, explore, and write out your ideas before going to the next page.

Right-Leg-Hypotenuse (RLH)

In particular, be sure to show why:

On the plane, ASS holds for right triangles (where the Angle in Angle-Side-Side is right).

This result is often called the *Right-Leg-Hypotenuse Theorem* (RLH), which can be expressed in the following way:

On the plane, if the leg and hypotenuse of one right triangle are congruent to the leg and hypotenuse of another right triangle, then the triangles are congruent.

At this point, you might conclude that RLH is true. for small triangles on a sphere. But there *are* small triangle counterexamples to RLH on spheres! The counterexample in Figure 7.9 will help you to see some ways in which spheres are intrinsically very different from the plane. We can see that the second leg of the triangle intersects the geodesic that contains the third side an infinite number of times. So on a sphere there are small triangles which satisfy the conditions of RLH although they are non-congruent.

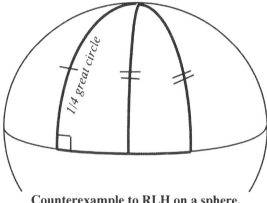

Figure 7.9   **Counterexample to RLH on a sphere.**

However, if you look at your argument for RLH on the plane, you should be able to show that

On a sphere, RLH is valid for a triangle with all sides less than 1/4 of a great circle.

RLH is also true for a much larger collection of triangles on a sphere. Can you find such a collection?

Chapter 8

Parallel Transport

> **Parallel** straight lines are straight lines which, being in the same plane and being produced indefinitely in both directions, do not meet one another in either direction.
>
>   — Euclid, *Elements*, Definition 23 [**A**: Euclid]

Problems 15, 16, and 17 allow us to develop further the notion of *parallel transport* that was introduced in Chapter 6.

PROBLEM 15. *Euclid's Exterior Angle Theorem (EEAT)*

> *Any exterior angle of a triangle is greater than each of the opposite interior angles.*[†]

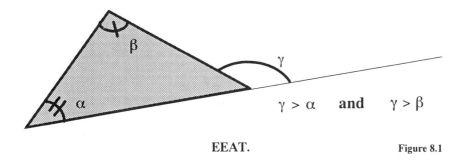

$$\gamma > \alpha \quad \text{and} \quad \gamma > \beta$$

EEAT.   **Figure 8.1**

Look at EEAT on both the plane and a sphere.

Suggestions

You may find the following hint (which is found in Euclid's writings) useful: Draw a line from the vertex of α to the midpoint, m, of the opposite side, bc. Extend that line beyond m to a point a' in such a

---

[†]**Warning**: Euclid's EAT is not the same as the Exterior Angle Theorem usually studied in high school.

way that $am \cong ma'$. Join a' to c. This hint will be referred to as *Euclid's hint*, and is pictured in Figure 8.2.

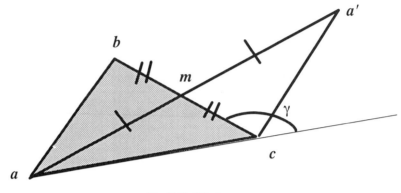

Figure 8.2   **Euclid's Hint.**

Be cautious transferring this hint to a sphere. It will probably help to draw Euclid's hint directly on a physical sphere.

It is not necessary to use Euclid's hint to prove EEAT, and in fact many people don't "see" the hint. Another perfectly good way to prove EEAT is to use Problem 16. Problems 15 and 16 are very closely related, and they can be done in either order. It is also fine to use 15 to prove 16 or use 16 to prove 15, but of course don't do both. As a final note, remember not to look at figures using only one orientation — rotations and reflections of a figure do not change its properties, so if you have trouble "seeing" something, check to see if it's something you're familiar with by orienting it differently on the page.

EEAT is not always true on a sphere, even for small triangles. Look at a counterexample as depicted in Figure 8.3. Then look at your proof of EEAT on the plane. It is very likely that your proof uses properties of angles and triangles that are true for small triangles on the sphere. Thus it may appear to you that your planar proof is also a valid proof of EEAT for small triangles on the sphere. But, there is a counterexample.

This could be, potentially, a very creative situation for you — **whenever you have a proof and counterexample of the same result, you have an opportunity to learn something deep and meaningful**. So, try out your planar proof of EEAT on the counterexample in Figure 8.3 and see what happens. Then try it on both large and small spherical triangles. If you can determine exactly which triangles satisfy EEAT and which triangles don't satisfy EEAT, then this information will be useful (but not crucial) to you in later problems.

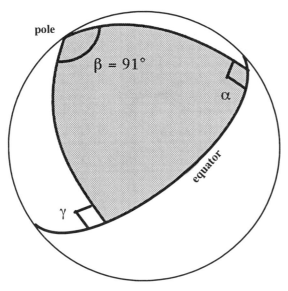

Counterexample to EEAT on a sphere.   Figure 8.3

PROBLEM 16. *Symmetries of Parallel Transported Lines*

Consider two lines, *r* and *r'*, that are parallel transports of each other along a third line, *l*. Consider now the geometric figure that is formed by the three lines, one of them being a transversal to the other two, and look for the symmetries of that geometric figure.

What can you say about the lines *r* and *r'*? Do they intersect? If so, where?

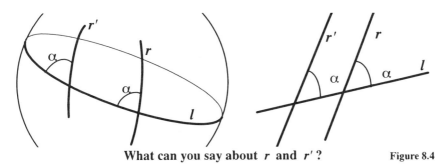

What can you say about *r* and *r'* ?   Figure 8.4

Suggestions

Parallel transport was already introduced informally in Chapter 6. In Problem 16 you have an opportunity to explore the concept further and prove its implications on the plane and a sphere. You will study the relationship between parallel transport and parallelism, as well.

A common high school definition of parallel lines is something like "two lines that never meet." But this is an inhuman definition — there is no way to check all points on both lines to see if they ever meet. This definition is also irrelevant on a sphere because we know that all geodesics on a sphere *will* cross each other. But we can measure the angles of a transversal. This is why it is more useful to talk about lines as parallel transports of one another rather than as parallel. So the question becomes:

> *If a transversal cuts two lines at congruent angles, are the lines in fact parallel in the sense of not intersecting?*

There are many ways to approach this problem. First, be sure to look at the symmetries of the local portion of the figure formed by the three lines. See what you can say about global symmetries from what you find locally. For the question of parallelism, you can use EEAT, but not if you used this problem to prove EEAT previously. Also, don't underestimate the power of symmetry when considering this problem. Many ideas that work on the plane will also be useful on a sphere, so try your planar proof on a sphere before attempting something completely different.

What is meant by *symmetry* with regard to geometric figures? A transformation is a *symmetry* of a geometric figure if it transforms that figure into itself. That is, the figure looks the same before and after the transformation. Here, we are looking for the symmetries of the plane and sphere.

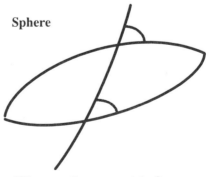

Sphere

Figure 8.5   **What are the symmetries?**

From this picture, we can see that on a sphere we are looking for the symmetries of a *lune* cut at congruent angles by a geodesic. A *lune* is a spherical region bounded by two half great circles.

You may be inclined to use one or both of the following results: *Any transversal of a pair of parallel lines cuts these lines at congruent angles* (Problem 20). And, *the angles of any triangle add up to a straight angle* (Problem 23). The use of these results should be avoided for now as they are both false on a sphere. We have been investigating what is common between the plane and spheres and trying to use common proofs whenever possible. In addition, we will find that it is necessary to make further assumptions about the plane before we can prove these results, and no additional assumptions are needed for Problem 16. You may be tempted to use other properties of parallel lines that seem familiar to you, but in each case ask yourself whether or not the property is true on a sphere. If it is not true on a sphere, then don't use it here since it is not needed.

PROBLEM 17. *Transversals through a Midpoint*

> *If two geodesics **r** and **r′** are parallel transports along another geodesic **l**, then they are also parallel transports along any geodesic passing through the midpoint of the segment of **l** between **r** and **r′**.*

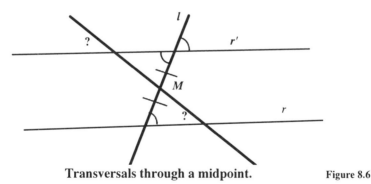

Transversals through a midpoint.   **Figure 8.6**

> *Does this work for both the plane and sphere? On a sphere are these the only lines which will cut **r** and **r′** at congruent angles? Why?*

This problem continues the ideas presented in Problem 16. In fact, you may have proven this problem while working on 16 without even knowing it. There are many ways to approach this problem. Using symmetry is always a good way to start. You can also use some of the triangle congruence theorems that you have been working with. On a sphere, look at the things you have discovered about transversals from Problems 15 and 16; they are very applicable here.

Since Chapter 6, you have been dealing with issues of parallelism. Parallel transport gives you a way to check parallelism. Even though parallel transported lines intersect on the sphere, there is a *feeling of local parallelness* about them. The issue is not whether the lines ever intersect but if they can be cut at congruent angles at certain points, that is, whether the lines are *locally parallel*. You may choose to avoid definitions of parallel that do not give you a direct method of verification. The following definitions are not human in the sense that they can not be verified directly:

◆   *Parallel lines are lines that never intersect*;

◆   *Parallel lines are lines such that any transversal cuts them at congruent angles*; or,

◆   *Parallel lines are lines that are everywhere equidistant.*

Problems 20 through 23 will force you to compare these various notions of parallelism and parallel transport, and, in the process, you may learn something about the history and philosophy of parallel lines.

On the plane, straight lines are parallel without ever intersecting; on a sphere, straight lines are locally parallel and converge symmetrically; and on a hyperbolic plane, straight lines can be locally parallel and diverge symmetrically (and thus not intersect). However, there are also pairs of "asymptotic" lines on the hyperbolic plane which do not intersect (and are thus parallel by the usual definition) but which are not parallel transports of each other along any transversal. If you look at a cone with cone angle larger than 360°, then you also can find two non-intersecting lines which are not parallel transports of each other. These examples should help you realize that parallelism is not just about non-intersecting lines.

Chapter 9
SAA and AAA

Things which coincide with one another are equal to one another.
— Euclid, *Elements*, Common Notion 4 [**A**: Euclid]

In Problems 18 and 19 you will investigate the two triangle congruence properties, "Side-Angle-Angle" and "Angle-Angle-Angle."

PROBLEM 18. *Side-Angle-Angle (SAA)*

Are two triangles congruent if one side, an adjacent angle, and the opposite angle of one triangle are congruent, respectively, to one side, an adjacent angle, and the opposite angle of the other triangle?

Suggestions

As a general strategy when investigating these problems, start by making the two triangles coincide as much as possible. You did this when investigating SSS and ASS. Let us try it as an initial step in our proof of SAA. Line up the first sides and the first angles. Since we don't know the length of the second side, we might end up with a picture like this:

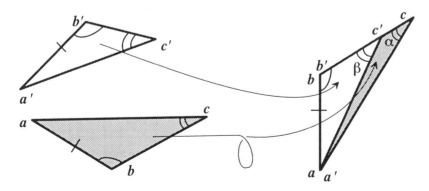

Starting SAA.   Figure 9.1

The situation shown in Figure 9.1 may seem to you to be impossible. You may be asking yourself, "Can this happen?" If your temptation is to argue that α and β cannot be congruent angles and that it is not possible to construct such a figure, behold Figure 9.2.

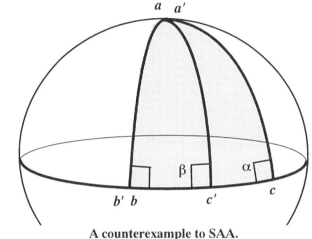

Figure 9.2   **A counterexample to SAA.**

You may be suspicious of this example because it is not a counterexample on the plane. You may feel certain that it is the only counterexample to SAA on a sphere. In fact, we can find other counterexamples for SAA on a sphere.

With the help of parallel transport, you can construct many counterexamples for SAA on a sphere. If you look back to the first counterexample given for SAA, you can see how this problem involves parallel transport, or similarly how it involves Euclid's Exterior Angle Theorem, which we looked at in Problem 15.

Can we make restrictions such that SAA *is* true on a sphere? You should be able to answer this question by using the fuller understanding of parallel transport you gained in Problems 15 and 16. You may be tempted to use the result, *the sum of the interior angles of a triangle is 180°*, in order to prove SAA on the plane. This result will be proven later (Problem 23) for the plane, but we say in Problem 8 that it does not hold on spheres. Thus, we encourage you to avoid using it and to use the concept of parallel transport instead. This suggestion stems from our desire to see what is common between the plane and sphere, as much as possible. In addition, before we can prove that the sum of the angles of a triangle is 180°, we will have to make some additional assumptions on the plane which are not needed for SAA.

PROBLEM *19. Angle-Angle-Angle (AAA)*

Are two triangles congruent if their corresponding angles are congruent?

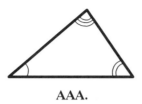

AAA.   **Figure 9.3**

As with the three previous problems, make the two AAA triangles coincide as much as possible. We know that we can line up one of the angles, but we don't know the lengths of either of the sides coming from this angle. So there are two possibilities: (1) Both sides of one are longer than both sides of the other, as this example shows on the plane,

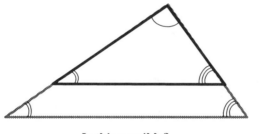

Is this possible?   **Figure 9.4**

or (2) One side of the first triangle is longer than the corresponding side of the second triangle and vice versa, as the example in Figure 9.5 shows on a sphere.

As with Problem 18, you may think that this example cannot happen on either a sphere or on the plane. The possible existence of a counterexample relies heavily on parallel transport — you can identify the parallel transports in each of the examples given. Try each counterexample on the plane and a sphere and see what happens. If these examples are not possible, explain why, and if they are possible, see if you can restrict the triangles sufficiently so that AAA does hold.

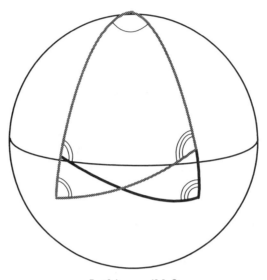

Figure 9.5   **Is this possible?**

Parallel transport shows up in AAA, similar to how it did in SAA, but here it happens simultaneously in two places. In this case, you will recognize that parallel transport produces similar triangles that are not necessarily congruent. However, there are no similar triangles on a sphere (as you will see after you finish AAA) and so you certainly need a proof that shows why such a construction is possible and why the triangles are not congruent. The construction may seem intuitively possible to you, but you should justify why it is a counterexample. Again you may need properties of parallel lines from Problems 15 through 17. You may also need the property of parallel transport on the plane stated in Problem 20 — you can assume this property now as long as you are sure not to use AAA when proving it later.

On a sphere, is it possible to make the two parallel transport constructions shown in Figure 9.5 and thus get two non-congruent triangles? Try it and see. It is important that you make such constructions and that you study them on a model of a sphere.

Chapter 10
Parallel Postulates

> Let the following be postulated: that, if a straight line falling on two straight lines make the interior angles on the same side less than two right angles, the two straight lines, if produced indefinitely, meet on that side on which are the angles less than the two right angles.
>
> — Euclid, *Elements*, Postulate 5 [**A**: Euclid]

Parallel Lines on the Plane Are Special

Up to this point we have not had to assume anything about parallel lines. No version of a parallel postulate has been necessary on either the plane or a sphere. We defined the concrete notion of parallel transport and proved in Problem 16 that on the plane parallel transported lines do not intersect. Now in this chapter we will look at two important properties on the plane:

> *If two lines on the plane are parallel transports of each other along some transversal, then they are parallel transports along any transversal.* (Problem 20)

> *On the plane the sum of the interior angles of a triangle is always* 180°. (Problem 23)

Neither of these properties is true on the sphere and both need an additional assumption on the plane for their proofs. The various assumptions that permit proofs of these two statements are collectively termed **Parallel Postulates**. Only the two statements above are needed from this chapter for the rest of the book. Therefore, it is possible to omit this chapter and assume one of the above two statements and then prove the other. However, parallel postulates have a historical importance and have a central position in many geometry textbooks and in many expositions about non-Euclidean geometries. The problems in this chapter are an attempt to help people unravel and enhance their understanding of parallel postulates. Comparing the situations on the plane and a sphere is a powerful tool for unearthing our hidden assumptions and misconceptions about the notion of parallel.

Since we have so many (often unconscious) connotations and assumptions attached to the word "parallel," we find it best to avoid using the term parallel as much as possible in this discussion. Instead we will use terms like "parallel transport," "non-intersecting," and "equidistant."

PROBLEM 20. Parallel Transport on the Plane

*Show that if l_1 and l_2 are lines on the plane such that they are parallel transports along a transversal l, then they are parallel transports along any transversal. Prove this using any assumptions you find necessary. Make as few assumptions as you can, and make them as simple as possible. **Be sure to state your assumptions clearly.***

What part of your proof does not work on a sphere?

Suggestions

This problem is by no means as trivial as it, at first, may appear. In order to prove this theorem, you will have to assume something — there are many possible assumptions, so use your imagination. But at the same time, try not to assume any more than is necessary. If you're having trouble deciding what to assume, try to solve the problem in a way that seems natural to you and see what develops.

On a sphere, try the same construction and proof you used for the plane. What happens? You should find that your proof does not work on a sphere. So, what is it about your proof (or the sphere) that creates difficulties?

Again, you may be tempted to use "the sum of the angles of a triangle is 180°" as part of your proof. As in many other cases before, there is nothing wrong with doing the problems out of order — you can use Problem 23 to prove Problem 20 as long as you don't also use Problem 20 to prove Problem 23. Most people find it much easier to prove Problem 20, first, and then use it to prove Problem 23.

Problem 20 emphasizes the differences between parallelism on the plane and parallelism on a sphere. On the plane, non-intersecting lines exist, and one can "parallel transport" everywhere. Yet, as was seen in Problems 16 and 17, on a sphere two lines are cut at congruent angles if and only if the transversal line goes through the center of the lune formed by them. That is, on a sphere two lines are locally parallel only when they can be parallel transported through the center of the lune formed by them. Be sure to draw a picture of the lune locating the center and the transversal. On a sphere it is impossible to slide the transversal along two parallel transported lines keeping both angles constant — which is something you can do on the plane. In Figure 10.1, the line t' is a parallel transport of line t along line l, but it is not a parallel transport of t along l'.

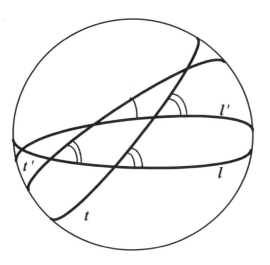

Parallel transport along l, but not along l'.   Figure 10.1

![pencil icon] **Pause, explore, and write out your ideas before going to the next page.**

Parallel Circles on a Sphere

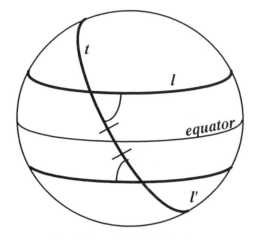

Figure 10.2   **Special equidistant circles.**

The latitude circles on the earth are sometimes called "Parallels of Latitude." They are parallel in the sense that they are everywhere equidistant as are concentric circles on the plane. In general, transversals do not cut equidistant circles at congruent angles. However, there is one important case where transversals do cut the circles at congruent angles. Let l and l' be latitude circles which are the same distance from the equator on opposite sides of it. See Figure 10.2. Then, every point on the equator is a center of half-turn symmetry for these pair of latitudes. Thus, as in Problems 17 and 20, every transversal cuts these latitude circles in congruent angles.

Parallel Postulates

One of Euclid's assumptions constitutes ***Euclid's Fifth (or Parallel) Postulate (EFP),*** which says:

> *If a straight line falling on two straight lines makes the interior angles on the same side less than two right angles, then the two straight lines, if produced indefinitely, meet on that side on which the angles are less than two right angles.*

For a picture of EFP, see Figure 10.3.

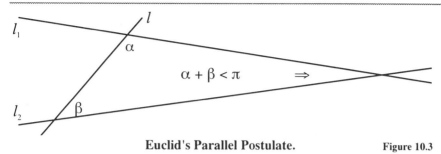

Euclid's Parallel Postulate.   Figure 10.3

You probably did not assume EFP in your proof of Problem 20. You are in good company — many mathematicians, including Euclid, have tried to avoid using it as much as possible. However, we will explore EFP because, historically, it is important, and because it has some very interesting properties as you will see in Problem 22. On a sphere, all straight lines intersect twice which means that EFP is trivially true on a sphere. But in Problem 22, you will show that EFP is also true in a stronger sense on spheres.

Thus EFP does not have to be assumed on a sphere — it can be proved! However, in most geometry text books, EFP is substituted by another postulate which, it is claimed, is equivalent to EFP. This postulate is ***Playfair's Parallel Postulate (PPP)***, and it can be expressed in the following way:

For every line l and every point P not on l, there is a unique line l′ which passes through P and is parallel to l.

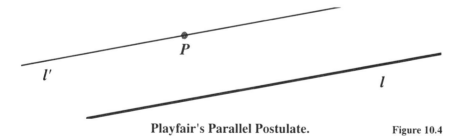

Playfair's Parallel Postulate.   Figure 10.4

Note that, on a sphere, since any two great circles intersect, there are *no* lines *l′* which are parallel to *l* in the "not intersecting" sense. Therefore, Playfair's Postulate is not true on spheres. On the other hand, if we change "parallel" to "parallel transport" then <u>every</u> great circle through *P* is a parallel transport of *l* along *some* transversal. In Problem 21, you will explore the relationships among EFP, Playfair's Postulate, and the assumptions you used in Problem 20.

PROBLEM *21. Parallel Postulates on the Plane*

*On the plane, are EFP, Playfair's Postulate, and your postu-
late from Problem 20 equivalent? Why? Or, why not?*

To show that EFP and Playfair's Postulate are equivalent on the
plane, you need to show that you can prove EFP if you assume Playfair's
Postulate and vice versa. Do the same for your postulate from Problem
20. If the three postulates are equivalent, then you can prove the equivalence by showing that

$$\text{EFP} \Rightarrow \text{PPP} \Rightarrow \text{Your Postulate} \Rightarrow \text{EFP}$$

or in any other order. It will probably help you to draw lots of pictures of
what is going on. Note that PPP is not true on a sphere but EFP is true, so
therefore your proof that EFP implies PPP on the plane must use some
property of the plane that does not hold on the sphere. Look for it.

PROBLEM *22. Parallel Postulates on a Sphere*

*(a) Show that EFP is true on the sphere in a strong sense;
that is, if lines l and l' are cut by a transversal t such that
the sum of the interior angles* $\alpha + \beta$ *on one side is less than
two right angles, then, not only do l and l' intersect, but they
also intersect "closest" to t on the side of* α *and* β*. You
will have to determine an appropriate meaning for "closest."*

*(b) Using the notion of parallel transport, change Playfair's
Postulate so that it is true on the sphere. Make as few alterations as possible and keep some form of uniqueness.*

*(c) Either prove your postulate from Problem 20 on the
sphere or change it, with as few alterations as possible, so that
it is true on the sphere.*

Suggestions for (a)

To help visualize the postulates, draw these "parallels" on an actual
sphere. There are really two parts to this proof — first, you must come up
with a definition of "closest" and, then, prove that EFP is true for this
definition. The two parts may come about simultaneously as you come
up with a proof. This problem is closely related to Euclid's Exterior Angle Theorem, but can also be proved without using EEAT. One case that
you should look at specifically is pictured in Figure 10.5. It is not necessarily obvious how to define the "closest" intersection.

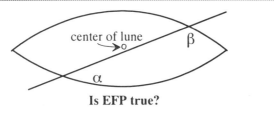

Is EFP true?   Figure 10.5

Suggestions for (b) and (c)

Above we noted that Playfair's Postulate is not true on the sphere, and in Problem 20 you should have decided whether or not your postulate is true on a sphere. The next step is to come up with a modified version of the postulates, one that *is* true on a sphere. Try to limit the modifications you make so that the new postulate preserves the spirit of the old one. You can draw ideas from any of the previous problems to obtain suitable modifications. Then, prove that your modified version of the postulate is true on a sphere.

Parallelism in Spherical and Hyperbolic Geometry

Playfair's Postulate requires more than that which is necessary to obtain parallel lines. In Problem 16, it was proven that if one line is a parallel transport of another, then the lines do not intersect on the plane; that is, they are parallel. On the sphere any two lines intersect. However, there *are* non-intersecting lines that are not parallel transports of each other on any cone with cone angle larger than 360°. (*Can you find an example on a 450° cone?*) The same situation exists on the hyperbolic plane which you can read about in other texts on non-Euclidean geometry. (See Section **K** in the Bibliography.) Hyperbolic geometry is a geometry that was discovered about 150 years ago by, C.F. Gauss (German), J. Bolyai (Hungarian), and N.I. Lobatchevsky (Ukrainian). Hyperbolic geometry is special from a formal axiomatic point of view because it satisfies all the axioms of Euclidean geometry except for the parallel postulate. Many written accounts talk about hyperbolic geometry and non-Euclidean geometry as being synonymous, but as we have seen there are many non-Euclidean geometries. It is also not accurate to say (as many books do) that non-Euclidean geometry was discovered about 150 years ago. Spherical geometry (which is clearly not Euclidean) was in existence and studied by at least the ancient Babylonians, Indians, and Greeks more than 2,000 years ago. Spherical geometry was of importance for astronomical observations and astrological calculations. Even Euclid in his *Phaenomena* [**A**: Euclid] (a work on astronomy) discusses propositions of spherical geometry. Menelaus, a Greek of the first century, published a book *Sphaerica* which contains many theorems about spherical triangles and compares them to triangles on the Euclidean plane. (*Sphaerica* survives only in an Arabic version. For a discussion see [**K**: Kline, page 120].)

Figure 10.6 is an attempt to represent the relationships among parallel transport, non-intersecting lines, EFP, and Playfair's Postulate. Can you fit your postulate into the diagram?

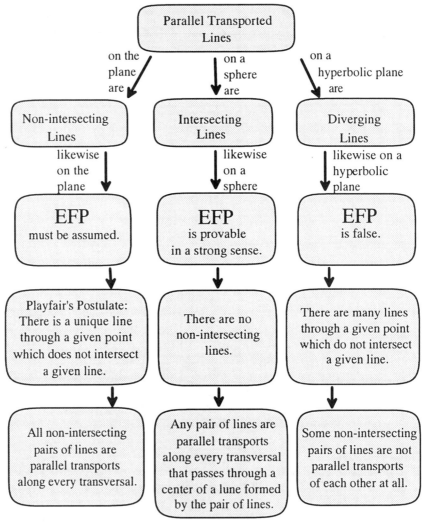

Figure 10.6   **Parallelism.**

PROBLEM 23. *Sum of the Angles of a Triangle*

> *What is the sum of the angles of a triangle on the plane and on a sphere?*

There are many approaches to this problem. It can be done using only results from this chapter or results from other chapters may also be used. Be sure to draw pictures and to be careful about what previous results you are using.

Remember: If you assumed facts about the sum of the angles of a triangle on the plane in a previous problem, then you can not use the results of that problem here.

What happens on the sphere?

Chapter 11
3-Spheres in 4-Space

> Let us, then, make a mental picture of our universe: ...as far as possible, a complete unity so that whatever comes into view, say the outer orb of the heavens, shall bring immediately with it the vision, on the one plane, of the sun and of all the stars with earth and sea and all living things as if exhibited upon a transparent globe.
>
> Bring this vision actually before your sight, so that there shall be in your mind the gleaming representation of a sphere, a picture holding all the things of the universe... Keep this sphere before you, and from it imagine another, a sphere stripped of magnitude and of spatial differences; cast out your inborn sense of Matter, taking care not merely to attenuate it: call on God, maker of the sphere whose image you now hold, and pray Him to enter. And may He come bringing His own Universe...
>
> — Plotinus, *The Enneads*, V.8.9 [A: Plotinus]

In this chapter you will explore the 3-dimensional sphere which extrinsically sits in 4-space. If we zoom in on a point in the 3-sphere, then intrinsically and locally the experience of the 3-sphere will become indistinguishable from an intrinsic and local experience of Euclidean 3-space. This is also our human experience in our physical universe. But, try to imagine the possibility of our physical universe being a 3-sphere in 4-space. It is the same kind of imagination a 2-dimensional being would need in order to imagine that it was on a plane or 2-sphere (ordinary sphere) in 3-space.

PROBLEM 24. Explain 3-Space to 2-Dimensional Person

How would you explain 3-space to a person living in two dimensions?

Think about the question in terms of this example: The person depicted in Figure 11.1 lives in a 2-dimensional plane. The person is wearing a mitten on the right hand. Notice that there is no front or back side to the mitten for the 2-D person. The mitten is just a thick line around the hand.

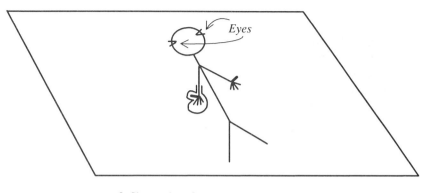

2-dimensional person with mitten.   Figure 11.1

Suppose that you approach the plane, remove the mitten, and put it on the 2-D person's left hand. There's no way within 2-space to move the mitten to fit the other hand. So, you take the mitten off of the 2-D plane, flip it over in 3-space, and then put it back on the plane around the left hand. The 2-D person has no experience of three dimensions but can see the result — the mitten disappears from the right hand, the mitten is gone for a moment, and then it is on the left hand.

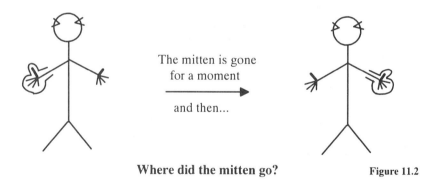

The mitten is gone
for a moment

and then...

Where did the mitten go?   Figure 11.2

How would you explain to the 2-D person what happened to the mitten?

Suggestions

This person's 2-dimensional experience is very much like the experience of an insect called a water strider that we talked about in Chapter 2. A water strider walks on the surface of a pond and has a very

2-dimensional perception of the universe around it. To the water strider, there is no up or down; its whole universe consists of the surface of the water. Similarly, for the 2-D person there is no front or back; the entire universe is the 2-dimensional plane.

Living in a 2-D world, the 2-D person can easily understand any figures in 2-space, including planes. In order to explain a notion such as "perpendicular," we could ask the 2-D person to think about the thumb and fingers on one hand.

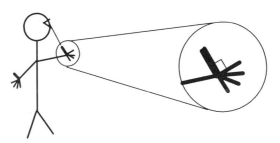

Figure 11.3   **The 2-D person sees "perpendicular."**

A person living in a 2-D world cannot directly experience three dimensions, just as we are unable to directly experience four dimensions. Yet, with some help from you, the 2-D person can begin to imagine three dimensions just as we can imagine four dimensions. One goal of this problem is to try to gain a better understanding of what our experience of 4-space might be. Think about what four dimensions might be like, and you may have ideas about the kinds of questions the 2-D person will have about three dimensions. You may know some answers, as well. The problem is finding a way to talk about them. Be creative!

One important thing to keep in mind is that it is possible to have *images* of things we cannot see. For example, when we look at a sphere, we can see only roughly half of it, but we can and do have an image of the entire sphere in our minds. We even have an image of the inside of the sphere, but it is impossible to actually see the entire inside or outside of the sphere all at once. Another similar example: sit in your room, close your eyes, and try to imagine the entire room. It is likely that you will have an image of the entire room, even though you can never see it all at once. Without such images of the whole room it would be difficult to maneuver around the room. The same goes for your image of the whole of the chair you are sitting on or this book you are reading.

Assume that the 2-D person also has images of things that cannot be seen in their entirety. For example, the 2-D person may have an image of a circle. Within a 2-dimensional world, the entire circle cannot be seen all at once; the 2-D person can only see approximately half of the outside of the circle at a time and can not see the inside at all unless the circle is broken.

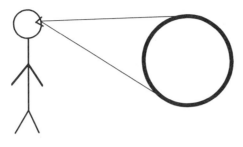

The 2-D person sees a circle.   Figure 11.4

However, from our position in 3-space we *can* see the entire circle including its inside. Carrying the distinction between what we can see and what we can imagine one step further, the 2-D person cannot see the entire circle but can imagine in the mind the whole circle including inside and out. Thus the 2-D person can only imagine what we, from three dimensions, can directly see. So, the 2-D person's image of the entire circle is as if it were being viewed from the third dimension. It makes sense, then, that the image of the entire sphere that we have in our minds is a 4-D view of it, as if we were viewing it from the fourth dimension.

When we talk about the fourth dimension here, we are not talking about time which is often considered the fourth dimension. Here, we are talking about a fourth *spatial* dimension. A fuller description of our universe would require the addition of a time dimension onto whatever spatial dimensions one is considering.

Try to come up with ways to help the 2-D person imagine what happens to the mitten when it is taken out of the plane into 3-space. It may help to think of intersecting planes rotating with respect to each other — How will a 2-D person in one of the planes experience it? Draw upon the person's experience living in two dimensions, as well as some of your own experiences and attempts to imagine four dimensions.

Terminology

We will now explore 3-dimensional spheres in 4-space which are possible in our physical universe. We include the following terminology to help clarify the terms and parameters of Problems 25-28. If you get bogged down in Problems 25-27, then jump to Problem 28.

- $\mathbb{E}^2$ = usual Euclidean plane.

- $\mathbb{E}^3$ = usual Euclidean 3-space:

  Every plane in $\mathbb{E}^3$ *has exactly one line perpendicular to it at every point.* [DEFINITION: A line is ***perpendicular*** to a plane if it intersects the plane and is perpendicular to every line in the plane which passes through the intersection point.]

- **2-sphere** (S^2) = points in $\mathbb{E}^3$ a fixed distance R from its center point C.

- $\mathbb{E}^4$ = Euclidean 4-space, in which:

  Every point p *on a 2-dimensional plane,* Π*, in* $\mathbb{E}^4$ *has an* ***orthogonal complement****,* $\Pi^{\perp}$*, which is a 2-dimensional plane which intersects* Π *only at* p *such that every line through* p *in* Π *is perpendicular to every line through* p *in* $\Pi^{\perp}$*.*

- **3-sphere** (S^3) = points in $\mathbb{E}^4$ a fixed distance R from its center point C.

  DEFINITION: We define a *great circle* on S^3 to be the intersection of S^3 with a plane in $\mathbb{E}^4$ through the center of the sphere, and we define a *great 2-sphere* on S^3 to be the intersection of S^3 with any 3-dimensional subspace of $\mathbb{E}^4$ that passes through the center of the sphere.

PROBLEM 25. *Intersecting Great Circles in the 3-Sphere*

If two great circles in S^3 *intersect, then they lie in the same great 2-sphere.*

Suggestions

Thinking in four dimensions may be a foreign concept to you, but believe it or not, it is possible to visualize a 4-dimensional space. Remember, the fourth dimension here is not time, but a fourth spatial dimension. We know that any two intersecting lines that are linearly independent (that do not coincide) determine a 2-dimensional plane. If we then add another line that is not in this plane, the three lines span a 3-space. When lines such as these are used as coordinate axes for a coordinate system, then they are typically taken to be orthogonal — each line is perpendicular to the others. Now to get 4-space, imagine a fourth line that is perpendicular to each of these original three. This creates the fourth dimension that we are considering.

Although we cannot experience all four dimensions at once, we can easily imagine any three at a time, and we can easily draw a picture of any two. This is the secret to looking at four dimensions. These 3- or 2-dimensional subspaces look exactly the same as any other 3-space or plane that you have seen before. This holds true for any subspace of 4-space — since all four of the coordinate lines are orthogonal, any set of three of these will look the same and will determine a space geometrically identical to our familiar 3-space, and any set of two coordinate lines will look like any other and will determine a 2-dimensional plane.

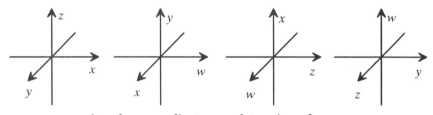

Any three coordinate axes determine a 3-space.   Figure 11.5

For all of these problems, you should not be looking at projections of the 3-sphere into a plane or a 3-space, but rather looking at the part of the 3-sphere that lies in a subspace. For example, since the 3-sphere is defined as the set of points a distance R from its center C in $\mathbb{E}^4$, if you take any 3-dimensional subspace of $\mathbb{E}^4$ through C, then the part of the 3-sphere which lies in this 3-dimensional subspace is the set of points a distance R from its center C in the 3-space. So any 3-dimensional subspace of $\mathbb{E}^4$ intersects the 3-sphere in a 2-sphere, which you know all about by now, and you can easily visualize.

For all of the problems here, it is generally best to draw pictures of various planes (2-dimensional subspaces) through the 3-sphere because they are easy to draw on a piece of paper. Remember, only include in your picture those geometric objects that lie in the plane you are drawing. So, a great circle that lies in this plane would be drawn as a circle, while another great circle that passed through this plane would intersect this plane only in two points. See Figure 11.6.

For this particular problem, you are looking at the 3-sphere extrinsically. A good way to proceed is to draw several planes as outlined above, and try to get an idea of how the planes relate to one another when combined into a 4-dimensional space. Once you have an understanding of how the different planes interact in four dimensions, it is fairly easy to show how the great circles of a 3-sphere behave.

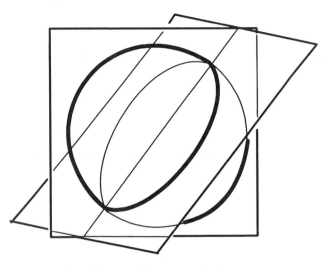

Figure 11.6   **Intersecting great circles.**

PROBLEM 26. Triangles in the 3-Sphere

Show that if A, B, C are three points in $\mathbf{S}^3$ that do not all lie on the same great circle, then there is a unique great 2-sphere, $\mathbf{G}^2$, containing A, B, C.

Thus we can define △ABC as the small triangle in $\mathbf{G}^2$ with vertices A, B, C. With this definition, **triangles in $\mathbf{S}^3$ have all the properties of small spherical triangles which we have been studying.**

Suggestions

Think back to the suggestions in Problem 25 — they will help you here, as well. Take two of the points, A and B,   and consider them together with the center, O, of the 3-sphere. These three points do not lie on the same line (*Why?*) and thus determine a plane and the intersection of this plane with the 3-sphere is a great circle. See Figure 11.6.

Think of A, B, and C as defining three intersecting great circles. Then look at the planes in which these great circles lie, and where the two planes lie in relation to one another.

Be sure to show that the great 2-sphere containing A,B,C is unique.

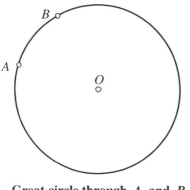

Great circle through *A* and *B*.   Figure 11.7

PROBLEM 27. *Disjoint Equidistant Great Circles*

Show that there are two great circles in S^3 *such that **every** point on one is a distance of one-fourth of a great circle away from **every** point on the other and vice versa.*

Suggestions

This problem is especially interesting because there is no equivalent theorem on the 2-sphere; we know that on the 2-sphere, all great circles intersect, so they can't be everywhere equidistant. The closest analogy on the 2-sphere is that a pole is everywhere equidistant from the equator. When we go up to the next dimension, this pole "expands" to a great circle such that every point on this great circle is everywhere equidistant from the equator. While this may seem mind-boggling, there are ways of seeing what is happening.

The main difference created by adding the fourth dimension lies in the orthogonal complement to a plane. In 3-space, the orthogonal complement of a plane is a line that passes through a given point. This means that for any given point on the plane, (the origin is always a convenient point), there is exactly one line that is perpendicular to the plane at that point. Now, what happens when you add the fourth dimension? In 4-space, the orthogonal complement to a plane is a plane. This means that every line in one plane is perpendicular to every line in the other plane. To understand how this is possible, think about how it works in 3-space and refer back to Figure 11.5. Now look at the *x-y*-plane and the *z-w*-plane. What do you notice? Why is every line through the center in one of these planes perpendicular to every line through the center in the other?

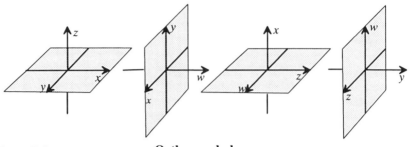

Figure 11.8   **Orthogonal planes.**

Knowing this, look at the two great circles in terms of the planes in which they lie, and look at the relationships between these two planes, i.e., where and how they intersect. Also, try to understand how great circles can be everywhere equidistant.

A Rotation That Moves Every Point

If we rotate along a great circle on a 2-sphere, all points of the sphere will move except for the two opposite poles of the great circle. If you rotate along a great circle on a 3-sphere, then the whole 3-sphere will move except for those points which are a quarter great circle away from the rotating great circle. Therefore, if you rotate along one of the two great circles which you found above, the other great circle will be left fixed. But now rotate the 3-sphere simultaneously along both great circles at the same speed. Now every point is moved!

(Optional) *Write an equation for this rotation and check that each point of the 3-sphere is moved at the same speed along some great circle.*

All of the great circles obtained by this rotation are equidistant from each other (in the sense that the perpendicular distance from every point on one great circle to another of the great circles is a constant). These great circles are traditionally called *Clifford parallels*. See [C: Penrose] for a readable discussion of Clifford parallels in the article entitled, *The Geometry of the Universe.*

Symmetries of Great Circles and Great 2-Spheres

We are now ready to see that the symmetries of great circles and great 2-spheres in a 3-space are the same as the symmetries of straight lines and (flat) planes in 3-space. If G is a great 2-sphere, then G is contained in a 3-dimensional subspace of 4-space. Now pick the center of the 3-sphere as the origin and define coordinates (x,y,z,w) on 4-space so that points in G have as their fourth coordinate $w = 0$. Then the transformation $(x,y,z,w) \longrightarrow (x,y,z,-w)$ that sends every point to a point with opposite fourth coordinate is a reflection of 4-space through the 3-space containing G, and it is a reflection of the 3-sphere through G. If g is a great circle in the 3-sphere, then let $g^{\perp}$ denote the great circle (from Problem 27) every point of which is $\pi/2$ from every point of g. Rotating along $g^{\perp}$ leaves g fixed and is a rotation about g, that is, a rotation with g as its axis. The following table gives a summary of various symmetries:

symmetries of...	reflection through...	reflection through...	half-turn about...	rotation about...	translation along...
line $l \subset R^2$	l	any line $\perp l$	any point in l	NA	l
great circle $g \subset S^2$	g	any great circle $\perp g$	any pt/pair in g	poles of g	g
line $l \subset R^3$	any plane $\supset l$	any plane $\perp l$	any line $\perp l$ and intersecting l	l	l
great circle $g \subset S^3$	any great sphere $\supset g$	any great sphere $\perp g$	any great circle $\perp g$ and intersecting g	g	g
plane $P \subset R^3$	P	any plane $\perp P$	any line in P	any line $\perp P$	any line $\subset P$
great circle $g \subset R^3$	$P(g) =$ plane $\supset g$	any 2-space $\perp P(g)$	any 1-space $\subset P(g)$	1-space $\perp P(g)$	NA
great sphere $G \subset S^3$	G	any great sphere $\perp G$	any great circle in G	any great circle $\perp G$	any great circle $\subset G$
plane $P \subset R^4$	any 3-space $\supset P$	any 3-space $\perp P$	any plane $\perp P$	P or $P^{\perp}$	any line $\subset P$
great circle $g \subset R^4$	any 3-space $\supset P(g)$	any 3-space $\perp P(g)$	any 2-space $\perp P(g)$	$P(g)$ or $P(g)^{\perp}$	NA

Look through the entries in the table and convince yourself that they are valid. After you have done this, you will be in a position to justify the claim that great circles in 3-spheres are intrinsically straight lines. Also, you should be able to show that, for any two points on a great 2-sphere, there is an intrinsically straight (with respect of the 3-sphere) path joining the two points that lies entirely in the great 2-sphere. (Such surfaces are often called *geodesically complete*.)

PROBLEM 28. *Is Our Universe a 3-Sphere?*

"What is the shape of our universe?" is still an open question. Some people have hypothesized that it is globally spherical. The current prevailing opinion among astrophysicists seems to be that the universe is globally hyperbolic. We say "globally" here in the sense that the earth is globally a sphere but has mountains and valleys which locally make the earth definitely not a sphere; in the same way, the physical universe is known to have local curvature near each star, such as our sun. But, the overall global shape has not yet been definitely determined.

The nineteenth century mathematician Carl Friedrich Gauss is said to have tried to measure the angles of a triangle whose vertices were three mountain peaks in Germany. If the sum of the angles had turned out to be other than 180°, then he would have surmised that the universe is not Euclidean. However, his measurements came to 180° within the accuracy of his measuring instruments. Could we do Gauss' experiment now, measuring the angles of a large triangle in our solar system?

Note that if the universe is a 3-dimensional sphere, the radius R of the universe would have to be at least as large as the diameter of our galaxy, which is about 6×10^{17} miles. In the foreseeable future, the largest triangle whose angles we could measure has area less than the area of our solar system, which is about 5×10^{19} square miles. So, if the universe were a 3-sphere, then we know that

$$\text{Area } (A) = [\Sigma \angle\text{'s} - \pi]\, R^2$$

(See Problems 25 and 40). So for a triangle the size of our solar system, the excess is given by

$$(\Sigma \angle\text{'s} - \pi) = A(\Delta)/R^2 < (5 \times 10^{19})/(6 \times 10^{17})^2 \approx 10^{-16} \text{ radians.}$$

This quantity is much too small to be measured, so, it is impractical to find out about the shape of our universe using the excess of a triangle within our solar system.

> *Given that it is impractical to measure the excess of triangles in our solar system, how might we get information as to whether the universe is Euclidean or spherical by only taking measurements of angles within our solar system but looking to the stars. [If you have trouble conceptualizing a 3-sphere, then*

you can do this problem for a very small bug on a 2-sphere who can see distant points (stars) on the 2-sphere, but who is restricted to staying inside its "solar system" which is so small that any triangle in it has excess too small to measure.]

Suggestions

Since we cannot possibly measure the three angles of a triangle in our solar system to the accuracy necessary to detect an excess, we must find other ways of determining the shape of our universe. Think back to everything you have learned about spheres and what it is that makes them different from space as Euclid described it. Always think intrinsically! You can assume, generally, that light will travel along geodesics, so think about looking at various objects and the relationships you would expect to find. For example, if we could see all the way around the universe (the distance of a great circle), then we could easily tell that the universe is spherical. Why? What if we could see half way around the universe? Or a quarter of the way around? Think of looking at stars at these distances.

Chapter 12
Dissection Theory

Oh, come with old Khayyám, and leave the Wise
To talk, one thing is certain, that Life flies;
One thing is certain, and the Rest is Lies;
The flower that once has blown for ever dies.
                                    — Omar Khayyam, *Rubaiyat*[†]

In showing that this parallelogram

Figure 12.1   **Parallelogram.**

has the same area as a rectangle with the same base and height (altitude),
we can easily cut the parallelogram into two pieces and rearrange them to
form this rectangle

Figure 12.2   **Equivalent by dissection to a rectangle.**

We say that two figures (*F* and *G*) are ***equivalent by dissection***
(*F* =$_d$ *G*) if one can be cut up into a finite number of pieces and the
pieces rearranged to form the other.

[†]From the translation by Edward Fitzgerald.

106

QUESTION: Is every plane figure equivalent by dissection to a square?

ANSWER: Yes! If the figure is a bounded polygon. There is a similar result that is true on the sphere. (Note that there are no squares on the sphere.)

You will prove these results about dissections in this chapter and the next and use them to look at the meaning of area. In this chapter you will show how to dissect any triangle or parallelogram into a rectangle with the same base. Then you will do analogous dissections on a sphere after first defining an appropriate analog of parallelograms and rectangles on the sphere. After that you will show that every polygon on the sphere is equivalent by dissection to a biangle and thus that two polygons which have the same area are equivalent by dissection to each other.

The proofs and solutions to all the problems can be done using "$=_d$", but if you wish you can use the weaker notion of "$=_s$": We say that two figures (F and G) are *equivalent by subtraction* ($F =_s G$) if there are two other figures, S and S', such that $S =_d S'$ and $F \cup S =_d G \cup S'$, where F & S and G & S' intersect at most in their boundaries. Saying two figures are equivalent by subtraction means that they can be arrived at by removing equivalent parts from two initially equivalent figures, like this:

Equivalent by subtraction.   **Figure 12.3**

If we cut out the two small squares as shown, we can see that the shaded portions of the rectangle and the parallelogram are equivalent by subtraction, but it is not at all obvious that one can be cut up and rearranged to form the other.

Equivalence by dissection is the stronger of the two methods of proof, and is generally preferable — it is much more obvious and in many ways, more convincing. All of the problems presented here can be proved by dissection, and we would urge you to use this method over equivalent by subtraction whenever you can.

Some of the dissection problems ahead are very simple, while some are rather difficult. If you think that a particular problem was so easy to solve that you may have missed something, chances are you hit the nail right on the head. For almost all of the problems, it is very helpful to make paper models and actually cut them up and fit the pieces together.

Most of the dissection proofs will consist of two parts: First show where to make the necessary cuts, and then prove that your construction works, i.e., that all the pieces do in fact fit together as you say they do.

PROBLEM 29. *Dissecting Plane Triangles*

> *Show that on the plane every triangle is equivalent by dis-section to a parallelogram with the same base no matter which base of the triangle you pick.*

Problem 29 is fairly straightforward, so don't try anything compli-cated. You only have to prove it for the plane — a proof for spheres will come in a later problem after we find out what to use in place of paral-lelograms, which do not exist on the sphere. Make paper models, and make sure your method works for all possible triangles with any side taken as the base. In particular, make sure that your proof works for triangles whose heights are much longer than their bases. Also, you need to show that the resulting figure actually is a parallelogram.

PROBLEM 30. *Dissecting Parallelograms*

> *Show that on a plane every parallelogram is equivalent by dissection to a rectangle with the same base and height.*

This problem is also rather simple, and a partial proof of this was given in the introduction at the beginning of this chapter. But for this problem, your proof must also work for tall, skinny parallelograms for which the given construction doesn't work:

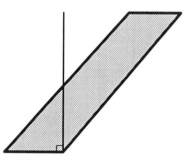

Figure 12.4   **Tall, skinny parallelogram.**

You may say that you can simply change the orientation of the par-allelogram, but, as for Problem 29, we want a proof that will work no matter which side you choose as the base. Again, don't try anything too complicated, and you only have to work on the plane.

Dissection Theory on Spheres

The above statements take on a different flavor when working on a sphere since one cannot construct parallelograms and rectangles, per se, on a sphere. We can define two types of polygons on a sphere, and then restate the above for spheres. The two types of polygons are the Khayyam quadrilateral and the Khayyam parallelogram. These definitions were first put forth by the Persian geometer-poet Omar Khayyam in the eleventh century AD.[†] Through a bit of Western chauvinism, geometry books generally refer to these quadrilaterals as Saccheri quadrilaterals after the Italian monk who translated into Latin and extended the works of Khayyam and others.

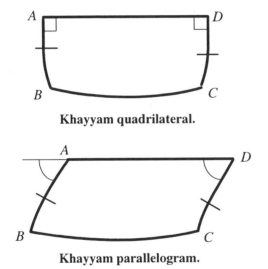

Khayyam quadrilateral.

Khayyam parallelogram.   Figure 12.5

A *Khayyam quadrilateral* (*KQ*) is a quadrilateral such that $AB \cong CD$ and $\angle BAD \cong \angle ADC \cong \pi/2$. A *Khayyam parallelogram* (*KP*) is a quadrilateral such that $AB \cong CD$ and AB is a parallel transport of DC along AD. In both cases, BC is called the *base* and the angles at its ends are called the *base angles*. One can check that a KQ on the plane is a rectangle and a KP on the plane is a parallelogram.

[†] See [**A**: Khayyam, 1931].

PROBLEM 31. *Khayyam Quadrilaterals*

Prove that the base angles of a KQ are congruent and that the perpendicular bisector of the top of a KQ is also the perpendicular bisector of the base.

To begin this problem, note that the definitions of KP and KQ make sense on the plane as well as on a sphere. The pictures in Figure 12.5 are deliberately drawn with a curved line for the base to emphasize the fact that the base angles are not necessarily congruent to the right angles. You should think of these quadrilaterals and parallelograms in terms of parallel transport instead of parallel lines. Everything you have learned about parallel transport and triangles on a sphere can be helpful for this problem. Symmetry can also be useful.

Now we are prepared to modify Problems 29 and 30 so that they will apply to spheres.

PROBLEM 32. *Dissecting Spherical Triangles*

Show that every small spherical triangle is equivalent by dissection to a Khayyam parallelogram with the same base.

Try your proof from Problem 29, as a first stab at the problem. You only need to look at a sphere. You should also look at the different proofs given for Problem 29. The only difference between the plane and sphere as far as this problem is concerned is that you must be more careful on a sphere because there are no parallel lines; there is only parallel transport. Some of the proofs for Problem 29 work well on a sphere, and others do not. Remember that the base of a KP is the side opposite the given congruent angles.

PROBLEM 33. *Dissecting Khayyam Parallelograms*

Prove that every Khayyam parallelogram is equivalent by dissection to a Khayyam quadrilateral with the same base.

As with Problem 32, start with your planar proof and work from there. As before, your method must work for tall, skinny KPs. Once you have come up with a construction, you must then show that the pieces actually fit together as you say they do, and prove that the angles at the top are right angles.

PROBLEM 34. *Spherical Polygons Dissect into Biangles*

In the next chapter you will show that every polygon on the plane is equivalent by dissection to a square. This does not apply to the sphere because there are no squares on the sphere. However, we have already shown in Problem 10 that two polygons on the same sphere have the same area if they have the same holonomy. Thus every polygon on the sphere must have the same area as the ***biangle*** (a region of the sphere bounded by two great circles, same as a lune) with the same holonomy. Now we can show that not only do they have the same area but they are also equivalent by dissection.

> *Show that every simple small polygon on a sphere is equivalent by dissection to a biangle with the same holonomy.*
> [That is, the angle of the biangle is equal to
>
> (½)(2π – sum of the exterior angles of the polygon).]
>
> *Consequently, two simple small polygons on the sphere with the same area are equivalent by dissection.*

Suggestions

The proof of this result can be completed by proving the following steps (or lemmas). (This proof was first suggested to me by my daughter, Becky.)

♦ *Every simple small polygon can be dissected into a finite number of small triangles, such that the holonomy of the polygon is the sum of the holonomies of the triangles.* [See Problems 10 and 11.]

♦ *Each small triangle is equivalent by dissection to a KQ with the same base and same holonomy.* [Check your proofs of Problems 32 and 33.]

♦ *Two KQ's with the same base and the same holonomy (or base angles) are congruent.*

♦ *If two Δ's have the same base and the same holonomy, then they are* $=_d$.

♦ *Any Δ is* $=_d$ *to a biangle with* H (Δ) = H (biangle) = *twice the angle of the biangle.*

Chapter 13
Square Roots, Pythagoras, and Similar Triangles

> The diagonal of an oblong produces by itself both the areas which the two sides of the oblong produce separately.
> — Baudhayana, *Sulbasutram*, Sutra 48 [A: Baudhayana]

In the last chapter, we showed that two polygons on a sphere with the same area are equivalent by dissection because both are equivalent to the same biangle. In this chapter, we will prove an analogous result on the plane by showing that every planar polygon is equivalent by dissection to a square.

In the process of exploring this dissection theory, we will follow a path through a corner of the forest of mathematics — a path that has delighted and surprised the author many times. We will bring with us the question: What are square roots? Along the way we will confront relationships between geometry and algebra of real numbers, in addition to similar triangles, the Pythagorean Theorem (the quote above is a statement of this theorem written before Pythagoras), and possibly the oldest written proof in geometry (at least 2,600 years old). This path will lead to the solutions of quadratic and cubic equations in Chapter 14. We will let the author's personal experience lead us on this path.

Square Roots

When I was in eighth grade, I asked my teacher, "What is a square root?" I knew that the square root of N was a number whose square was equal to N, but where would I find it? (Hidden in that question is "How do I know it always exists?") I knew what the square roots of 4 and 9 were — no problem there. I even knew that the square root of 2 was the length of the diagonal of a unit square, but what of the square root of 2.5 or of π?

At first, the teacher showed me a square root table (a table of numerical square roots), but I soon discovered that if I took the number listed in the table as the square root of 2 and squared it, I got 1.999396, not 2. (Modern-day pocket calculators give rise to the same problem.)

112

So I persisted in asking my question — What is the square root? Then the teacher answered by giving me THE ANSWER — the Square Root Algorithm. Do you remember the Square Root Algorithm — that procedure, similar to long division, by which it is possible to calculate the square root? Or perhaps more recently you were taught the "divide and average" method which goes like this:

> If A_1 is an approximation of the square root of N, then the average of A_1 and N/A_1 is an even better approximation which we could call A_2. And then the next approximation A_3 is the average of A_2 and N/A_2. In equation form this becomes

$$A_{n+1} = (1/2)(A_n + (N/A_n)).$$

> For example, if $A_1 = 1.5$ is an approximation of the square root of 2, then

$$A_2 = 1.417\cdots, \quad A_3 = 1.414216\cdots$$

and so forth are better and better approximations.

But wait! Most of the time these algorithms do not calculate the square root — they only calculate approximations to the square root. The algorithms have an advantage over the tables because I could, at least in theory, calculate approximations as close as I wished. However, they are still only approximations and my question still remained — What is the square root which these algorithms approximate?

My eighth-grade teacher then gave up, but later in college I found out that some modern mathematicians answer my question in the following way: "We make an assumption (the Completeness Axiom) which implies that the sequence of approximations from the Square Root Algorithm must converge to some real number." And, when I continued to ask my question, I found that in modern mathematics the square root is a certain equivalence class of Cauchy sequences of rational numbers, or a certain Dedekind cut. Finally, I let go of my question and forgot it in the turmoil of graduate school, writing my thesis and beginning my mathematical career.

Later, I started teaching the geometry course that is the basis for this book. One of the problems in the course is the following problem.

PROBLEM 35. A Rectangle Dissects into a Square

Show that, on the plane, every rectangle is equivalent by dissection to a square.

Suggestions

Problems 35 and 36 can be done in any order. So if you get stuck on one problem, you can still go on to the other. In Problems 35 and 36, it is especially important to make accurate models and constructions — rough drawings will not show the necessary length and angle relationships.

Problem 35 is one of the oldest problems in geometry, so you may have guessed (correctly!) that it is one of the more complex ones. This problem is interesting for more than just historical reasons. You are asked to prove that you can cut up any rectangle (into a finite number of pieces) and rearrange the pieces to form a square, like so:

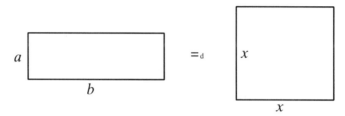

Figure 13.1   **Dissecting a rectangle into a square.**

Since you are neither adding anything to the rectangle nor removing anything, the area must remain the same, so $ab = x^2$, or $x = \sqrt{ab}$. What you are really finding is a geometric interpretation of a square root.

Let us look at a proof which is similar to the proofs in many standard geometry textbooks:

> Let $s = \sqrt{ab}$ be the side of the square equivalent by dissection to the rectangle with sides a and b. Place the square , *AEFH*, on the rectangle, *ABCD*, as shown in Figure 13.2. Draw *ED* to intersect *BC* in *R* and *HF* in *K*. Let *BC* intersect *HF* in *G*.[†] From the similar triangles $\triangle KDH$ and $\triangle EDA$ we have $HK/AE = HD/AD$, or
>
> $$HK = (AE)(HD)/AD = s(a - s)/a = s - s^2/a = s - b.$$
>
> Therefore, we have $\triangle\ EFK \cong \triangle RCD,\ \triangle EBR \cong \triangle KHD$.

[†] In the case that *ABCD* is so long and skinny that *K* ends up between *G* and *F*, we can, by cutting *ABCD* in half and stacking the halves, reduce the proof to the case above.

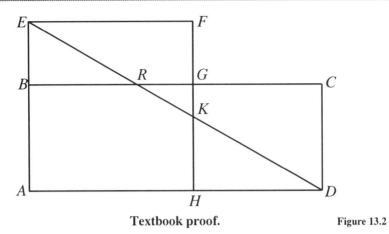

Textbook proof.   Figure 13.2

I was satisfied with the proof until in the second year of the course I started sensing student uneasiness with it. As I listened to their comments, I noticed questions being asked by the students: "What is $\sqrt{ab}$?" "How do you find it?" Those used to be my questions!

The students and I also noticed that the facts used about similar triangles in the proof above are usually proved using the theory of areas of triangles. Thus, this proof could not be used as part of a concrete theory of areas of polygons, which was our purpose in studying dissection theory in the first place. Notice that the above proof also assumes the existence of the square which in analysis is based on the Completeness Axiom. The conclusion here seems to be that it would be desirable instead to construct the square root x. That started me on an exploration which continued on and off for many years.

Now let us solve a few problems in dissection theory:

Here are three methods for constructing x. For all three constructions we will use a rectangle like the one shown in Figure 13.1, with the longer side b as the base and the shorter side a as the height.

For the first construction, Figure 13.3, take the rectangle and lay a out to the left of b. Use this base line as the diameter of a circle. The length x that you're looking for is the perpendicular line from the left side of the triangle to where it intersects the circle:

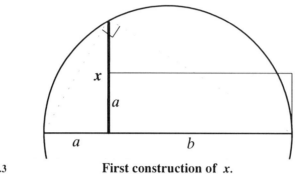

Figure 13.3   **First construction of** *x*.

The second construction, Figure 13.4, is similar but this time put *a* on the inside of the base of the rectangle. Now the side *x* you are looking for is the segment from the lower left corner of the rectangle to the point at which a perpendicular rising from the place you put *a* intersects the circle:

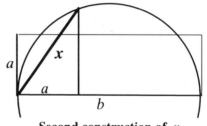

Figure 13.4   **Second construction of** *x*.

The third construction is a bit more algebraic than the others, and doesn't directly involve a circle:

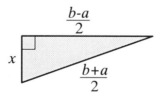

Figure 13.5   **Third construction of** *x*.

This construction can be used together with the result of Problem 36 in order to obtain a proof by subtraction.

For all of these constructions, it is imperative that you use accurate models. Whatever method you choose, make the rectangle and the square overlap as much as possible, and see how to fit the other pieces in. Then, you have to prove that all of the sides and angles line up properly. Note that it is much better to solve this problem geometrically rather than by only trying to work out the algebra — actually do the construction and proceed from there. Finally, you don't have to use one of the constructions shown here. If these don't make sense to you, then find one of your own that does.

And remember not to use results about similar triangles since we normally need results about areas in order to prove these results as we will do in Problem 37.

Pause, explore, and write out your ideas before going to the next page.

Baudhayana's Sulbasutram

While reading an unrelated article, I ran across an item that said that the problem of changing a rectangle into a square appeared in the *Sulbasutram* [**A**: Baudhayana]. "Sulbasutram" means "rules of the cord" and is an ancient (at least 600 BC) Sanskrit book written as a manual for people who were building temples in ancient India. Most of the book gives detailed instructions on temple and altar construction and design, but the first chapter is a geometry textbook which contains geometric statements called "Sutra." Sutra 54 is what we asked you to prove in Problem 35. It states:

> If you wish to turn an oblong[†] into a square, take the shorter side of the oblong for the side of square. Divide the remainder into two parts and inverting join those two parts to two sides of the square. Fill the empty place by adding a piece. It has been taught how to deduct it.

Here is a diagram for Sutra 54:

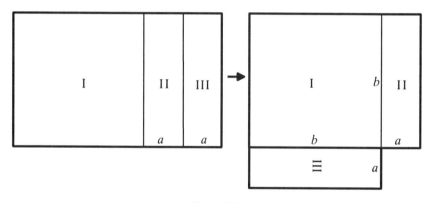

Figure 13.6   **Sutra 54.**

So our rectangle has been changed into a figure with an "empty place" which can be filled "by adding a piece" (a small square). The result is a large square from which a small square has to be removed (or "deducted").

Now Sutra 51:

> If you wish to deduct one square from another square, cut off a piece from the larger square by making a mark on the ground with the

---

[†] Here "oblong" means "rectangle."

side of the smaller square which you wish to deduct; draw one of the sides across the oblong so that it touches the other side; by this line which has been cut off the small square is deducted from the large one.

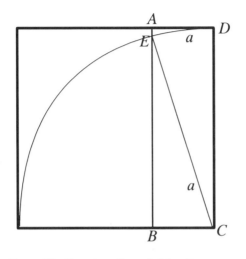

Sutra 51: Construction of side of square.   Figure 13.7

We wish to deduct the small square (a^2) from the large square. Sutra 51 tells us to "scratch up" with the side of the smaller square — this produces the line *AB* and the oblong *ABCD*. Now, if we "draw" the side *CD* of the large square to produce an arc, then this arc intersects the other side at the point *E*. The sutra then claims that *BE* is side of the desired square whose area equals the area of the large square minus the area of the small square. This last assertion follows from Sutra 50, which we ask you to prove in Problem 36.

See Figure 13.8 for the drawing that goes with Sutra 50:

If you wish to combine two squares of different size into one, scratch up with the side of the smaller square a piece cut off from the larger one. The diagonal of this cutoff piece is the side of the combined squares.

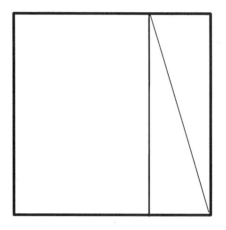

Figure 13.8   **Sutra 50.**

Be sure you see why Sutra 50 is a statement of what we call the Pythagorean Theorem.

A. Seidenberg, in an article entitled *The Ritual Origin of Geometry* [**G**: Seidenberg], gives a detailed discussion of the significance of the Sulbasutram. He argues that it was written before 600 BC (Pythagoras lived about 500 BC and Euclid about 300 BC). He gives evidence to support his claim that it contains codification of knowledge going "far back of 1700 BC" and that knowledge of this kind was the common source of Indian, Egyptian, Babylonian, and Greek mathematics. Together Sutras 50, 51, and 54 describe a construction of a square with the same area as a given rectangle (oblong) and a proof (based on the Pythagorean Theorem) that this construction is correct. You can find stated in many books and articles that the ancient Hindus, in general, and the Sulbasutram, in particular, did not have proofs or demonstrations, or they are dismissed as being "rare." However, there are several sutras in [**A**: Baudhayana] similar to the ones discussed above. I suggest you decide for yourself to what extent they constitute proofs or demonstrations.

Baudhayana avoids the Completeness Axiom by giving an explicit construction of the side of the square. The construction can be summarized in Figure 13.9:

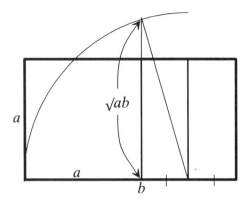

Baudhayana's construction of square root.   Figure 13.9

This is the same as Euclid's construction in Proposition II-14 (see [A: Euclid, page 409]). But Euclid's proof is much more complicated. Note that neither Baudhayana nor Euclid gives a proof of Problem 35 because the use of the Pythagorean Theorem obscures the dissection. However, they do give a concrete construction and a proof that the construction works. Both Baudhayana and Euclid prove the following theorem which uses equivalence by subtraction.

> **THEOREM:** *For every rectangle R there are squares S_1 and S_2 such that $R + S_2$ is equivalent by dissection to $S_1 + S_2$ and thus R and S_1 have the same area.*

Notice that both Baudhayana's and Euclid's proofs of this theorem and your proof of Problem 35 avoid assuming that the square root exists (and thus avoid the Completeness Axiom). They also avoid using any facts about similar triangles. These proofs explicitly construct the square and show in an elementary way that its area is the same as the area of the rectangle. There is no need for the area or the sides of the rectangle to be expressed in numbers. Also given a real number, b, the square root of b can be constructed by using a rectangle with sides b and 1. In Problem 37 we will use these techniques to prove basic properties about similar triangles.

So, finally, we have an answer to our question — What is a square root? It is "**an** answer" because many other solutions given from different points of view may be on the horizon.

PROBLEM 36. *Equivalence of Squares*

Prove the following: On the plane, the union of two squares is equivalent by dissection to another square.

Suggestions

This is closely related to the Pythagorean Theorem. There are two general ways to approach this problem: You can use Problem 35 or you can prove it on its own, which will result in a proof of the Pythagorean Theorem — hence, you can't use the Pythagorean Theorem to solve this problem because you will be proving it!

To see how this problem relates to the Pythagorean Theorem, think about the following statement of the Pythagorean Theorem:

The square on the hypotenuse is equal to the sum of the squares on the other two sides.

This is not just an algebraic equation — the squares referred to are actual geometric squares.

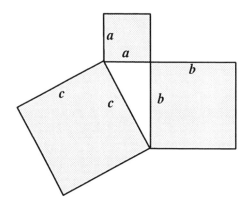

Figure 13.10   **Pythagorean Theorem.**

The term "completing the square" likewise came from geometry, and has some application in this problem as well as in Problem 35. As in the other dissection problems, make the three squares coincide as much as possible, and see how you might get the remaining pieces to overlap. You might start by reflecting the square with side *c* over the side *c* on the triangle. Then prove that the construction works as you say it does.

Any Polygon Can Be Dissected into a Square

If you put together Problems 29, 30, 35, and 36, a surprising result is created:

> *On the plane, every polygon is equivalent by dissection to a square.*

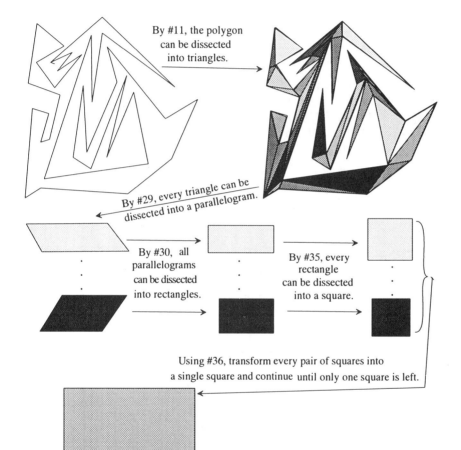

By #11, the polygon can be dissected into triangles.

By #29, every triangle can be dissected into a parallelogram.

By #30, all parallelograms can be dissected into rectangles.

By #35, every rectangle can be dissected into a square.

Using #36, transform every pair of squares into a single square and continue until only one square is left.

Every polygon dissects into a square.　　　　**Figure 13.11**

PROBLEM 37. Similar Triangles

Near the beginning of this chapter we gave a textbook proof of Problem 35 which used properties of similar triangles. Later you found a proof that did not need to use similar triangles. Now you are ready to give a dissection proof of:

> *If two triangles have corresponding angles congruent, then the corresponding sides of the triangles are in the same proportion to one another.*

Suggestions

Look at your proof of Problem 35. It probably shows implicitly that Problem 37 holds for a pair of similar triangles in your construction. For more generality, let θ be one of the angles of the triangles and place the two θ's in VAT position in such a way that they form two parallelograms as Figure 13.12.

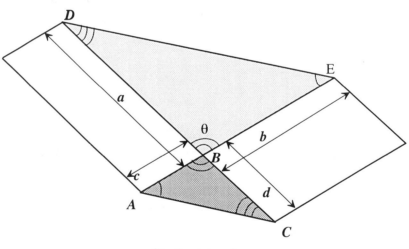

Figure 13.12   **Similar triangles.**

Show that the two parallelograms are equivalent by dissection and use that result to show that $ac = bd$ or in equivalent form $a/d = c/b$. You may find it clearer if you start by looking at the special case of $\theta = \pi/2$.

Three-Dimensional Dissections

In 1900, David Hilbert delivered a lecture before the International Congress of Mathematicians in which he listed 23 problems "from the discussion of which an advancement of science may be expected." These problems are now called ***Hilbert's Problems***.

Hilbert's Third Problem[†] asks whether it is possible to specify

> *two tetrahedra of equal bases and equal altitudes which can in no way be split up into congruent tetrahedra, and which cannot be combined with congruent tetrahedra to form two polyhedra which themselves could not be split up into congruent tetrahedra.*

Shortly after Hilbert's lecture, Max Dehn found such tetrahedra and also proved that a regular tetrahedron is not equivalent by dissection to a cube. Thus there is no possibility of dissecting polyhedra into cubes. To show these results, Dehn proved the following theorem:

> *If P and Q are two polyhedra in 3-space that are equivalent by dissection (or by subtraction), then the dihedral angles (see Chapter 18) of P are, mod π, a rational linear combination of the dihedral angles of Q. I.e., if α_i are the dihedral angles of P and β_j, are the dihedral angles of Q, then there are integers n_i, m_j, a, b such that*

$$\sum_i n_i \alpha_i + a\pi = \sum_j m_j \beta_j + b\pi.$$

[†] See [**Q**: Sah] for a discussion of this problem and its history.

Chapter 14

Geometric Solutions of Quadratic and Cubic Equations

> Whoever thinks algebra is a trick in obtaining unknowns has thought it in vain. No attention should be paid to the fact that algebra and geometry are different in appearance. Algebras (jabbre and maqabeleh) are geometric facts which are proved by Propositions Five and Six of Book Two of [Euclid's] Elements.
> — Omar Khayyam, a paper [A: Khayyam (1963)]

In this chapter we will see how the results from Chapter 13 were used historically to solve equations. Quadratic equations were solved by "completing the square" — a real square. These in turn lead to conic sections and cube roots and culminate in the beautiful general method from Omar Khayyam which can be used to find all the real roots of cubic equations. Along the way we shall clearly see some of the ancestral forms of our modern Cartesian coordinates and analytic geometry. I will point out several inaccuracies and misconceptions that have crept into the modern historical accounts of these matters. But I urge you to not look at this only for its historical interest but rather also look for the meaning it has in our present-day understanding of mathematics. This path is not through a dead museum or petrified forest; it passes through ideas which are very much alive and which have something to say to our modern technological, increasingly numerical, world.

PROBLEM 38. Quadratic Equations

Finding square roots is the simplest case of solving quadratic equations. If you look in some history of mathematics books (e.g., [G: Joseph] and [G: Eves]), you will find that quadratic equations were extensively solved by the Babylonians, Chinese, Indians, and Greeks. However, the earliest known general discussion of quadratic equations took place between 800 and 1100 AD in the Muslim Empire. Best known are Mohammed Ibn Musa al'Khowarizmi (who lived in Baghdad and from whose name we get our word "algorithm") and Omar Khayyam (the Persian geometer who is mostly known in the West for his philosophical poetry *The Rubaiyat*). Both wrote books entitled *Al-jabr w'al mugabalah* (from which we get our word "algebra"), al'Khowarizmi in about 820 AD and Khayyam in about 1100 AD. An English translation of both books is available in many libraries, if you can figure out whose name it is catalogued under (see, [**A:** al'Khowarizmi] and [**A:** Khayyam, 1931]). We previously met Khayyam in Chapter 12.

In these books you find geometric and numerical solutions to quadratic equations and geometric proofs of these solutions. But the first thing that you notice is that there is not one general quadratic equation as we are used to it:

$$ax^2 + bx + c = 0.$$

Rather, because the use of negative coefficients and negative roots was avoided, they list six types of quadratic equations (we follow Khayyam's lead and set the coefficient of x^2 equal to 1):

1. $bx = c$, which needs no solution,

2. $x^2 = bx$, which is easily solved,

3. $x^2 = c$, which has root $x = \sqrt{c}$,

4. $x^2 + bx = c$, with root $x = \sqrt{(b/2)^2 + c} - b/2$,

5. $x^2 + c = bx$, with roots $x = b/2 \pm \sqrt{(b/2)^2 - c}$,   if $c <$ $(b/2)^2$, and

6. $x^2 = bx + c$, with root $x = b/2 + \sqrt{(b/2)^2 + c}$.

Here b and c are always positive numbers or a geometric length (b) and area (c).

a. *Show that these are the only types. Why is $x^2 + bx + c = 0$ not included? Explain why b must be a length but c an area.*

The avoidance of negative numbers was widespread until a few hundred years ago. In the sixteenth century, European mathematicians called the negative numbers that appeared as roots of equations "numeri fictici" — fictitious numbers (see [**A:** Cardano, page 11]). In 1759 Baron Francis Masères, mathematician and a Fellow at Cambridge University and a member of the Royal Society, wrote in his *Dissertation on the Use of the Negative Sign in Algebra*:

> ...[negative roots] serve only, as far as I am able to judge, to puzzle the whole doctrine of equations, and to render obscure and

mysterious things that are in their own nature exceeding plain and simple.... It were to be wished therefore that negative roots had never been admitted into algebra or were again discarded from it: for if this were done, there is good reason to imagine, the objections which many leaned and ingenious men now make to algebraic computations, as being obscure and perplexed with almost unintelligible notions, would be thereby removed; it being certain that Algebra, or universal arithemtic, is, in its own nature, a scinece no less simple, clear, and capable of demonstration, than geometry.

More recently in 1831, Augustus De Morgan, a famous professor of mathematics at University College, London, wrote in his *On the Study and Difficulties of Mathematics*:

The imaginary expression $\sqrt{-a}$ and the negative expression $-b$ have this resemblance, that either of them occurring as solution of a problem indicates some inconsistency or absurdity. As far as real meaning is concerned, both are equally imaginary, since $0 - a$ is as inconceivable as $\sqrt{-a}$.

b. *Speculate about why these mathematicians avoided negative numbers and why they said what they said?*

To get a feeling for why, think about the meaning of 2×3 as two 3's and 3×2 as three 2's and then try to find a meaning for $3 \times (-2)$ and $-2 \times (+3)$. Also consider the quotation at the beginning of this chapter from Omar Khayyam about algebra and geometry. Some historians have quoted this passage but have left out all the words appearing after "proved." In my opinion, this omission changes the meaning of the passage. Euclid's propositions that are mentioned by Khayyam are the basic ingredients of Euclid's proof of the square root construction and form a basis for the construction of conic sections — see below. Geometric justification when there are negative coefficients is at the least very cumbersome, if not impossible. (If you doubt this, try to modify some of the geometric justifications below.)

c. *Find geometrically the algebraic equations which express all the positive roots of each of the six types. Fill in the details in the following sketch of Khayyam's methods for Types 3-6.*

For the geometric justification of Type 3 and the finding of square roots, Khayyam refers to Euclid's construction of the square root in Proposition II 14, which we discussed in Chapter 13, Problem 35.

For Type 4, Khayyam gives the following as geometric justification:

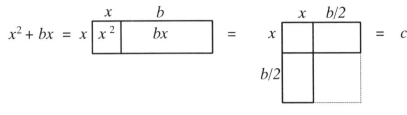

$$x^2 + bx = x \begin{array}{|c|c|} \hline x^2 & bx \\ \hline \end{array} = c$$

Type 4.   Figure 14.1

and thus, by "completing the square" on $x + b/2$, we have
$$(x + b/2)^2 = c + (b/2)^2.$$

Thus we have $x = \ldots$? Note the similarity between this and Baudhay-ana's construction of the square root (see Chapter 13).

For Type 5, Khayyam first assumes $x < (b/2)$ and draws the equation as:

$$x^2 + c = x \begin{array}{|c|c|} \hline x^2 & c \\ \hline \end{array} = bx$$

Type 5, $x < (b/2)$.   Figure 14.2

and note that the square on $b/2$ is $(b/2 - x)^2 + c.$

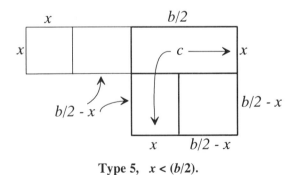

Type 5, $x < (b/2)$.   Figure 14.3

This leads to $x = b/2 + \sqrt{(b/2)^2 + c}$. Note that if $c > (b/2)^2$, then this geometric solution is impossible. When $x > (b/2)$, Khayyam uses the drawings:

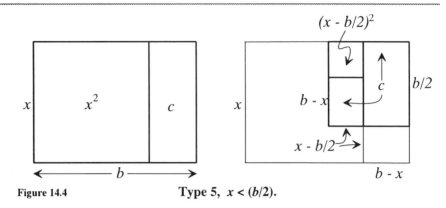

Figure 14.4   **Type 5, $x < (b/2)$.**

For solutions of Type 6, Khayyam uses the drawing in Figure 14.5.

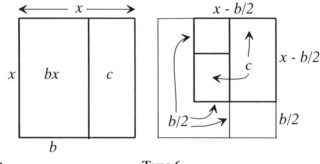

Figure 14.5   **Type 6.**

Do the above solutions find the negative roots? Well, first, the answer is clearly, No, if you mean: Did al'Khowarizmi and Khayyam mention negative roots? But let us not be too hasty. Suppose $-r$ (r, positive) is the negative root of $x^2 + bx = c$. Then $(-r)^2 + b(-r) = c$ or $r^2 = br + c$. Thus r is a positive root of $x^2 = bx + c$! The absolute value of the negative root of $x^2 + bx = c$ is the positive root of $x^2 = bx + c$ and vice versa. Also, the absolute values of the negative roots of $x^2 + bx + c = 0$ are the positive roots of $x^2 + c = bx$. So, in this sense, *Yes, the above geometric solutions do find all the real roots of all quadratic equations.* Thus it is misleading to state, as most historical accounts do, that the geometric methods failed to find the negative roots. The users of these methods did not find negative roots because they did not conceive of them. However, the methods can be directly used to find all the positive and negative roots of all quadratics.

    d. *Use Khayyam's methods to find all roots of the following equations:* $x^2 + 2x = 2$, $x^2 = 2x + 2$, $x^2 + 3x + 1 = 0$.

PROBLEM 39. *Conic Sections and Cube Roots*

The Greeks noticed that, if $a/c = c/d = d/b$, then $(a/c)^2 = (c/d)(d/b) = (c/b)$ and thus $c^3 = a^2 b$. Now setting $a = 1$, we see that we can find the cube root of b, if we can find c and d such that $c^2 = d$ and $d^2 = bc$. If we think of c and d as being variables and b a constant, then we see these equations as the equations of two parabolas with perpendicular axes and the same vertex. The Greeks also saw it this way but first they had to develop the concept of a parabola!

To the Greeks, and later Khayyam, if AB is a line segment, then *the parabola with vertex B and parameter AB* is the curve P such that, if C is on P, then the rectangle $BDCE$ (see Figure 14.6) has the property that $(BE)^2 = BD \cdot AB$. Since in Cartesian coordinates the coordinates of C are (BE, BD) this last equation becomes a familiar equation for a parabola.

Points of the parabola may be constructed by using the construction for the square root given in Chapter 13. In particular, E is the intersection of the semicircle on AD with the line perpendicular to AB at B. (The construction can also be done by finding D' such that $AB = DD'$, then the semicircle on BD' intersects P at C.) I encourage you to try this construction yourself; it is very easy to do if you use a compass and graph paper.

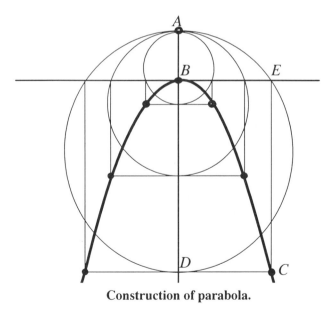

Construction of parabola.   Figure 14.6

Now we can find the cube root. Let b be a positive number or length and let $AB = b$ and construct C so that CB is perpendicular to AB and such that $CB = 1$. See Figure 14.7. Construct a parabola with vertex B and parameter AB and construct another parabola with vertex B and parameter CB. Let E be the intersection of the two parabolas. Draw the rectangle $BGEF$. Then

$$(EF)^2 = BF{\cdot}AB \text{ and } (GE)^2 = GB{\cdot}CB.$$

But, setting $c = GE = BF$ and $d = GB = EF$, we have

$$d^2 = cb \text{ and } c^2 = d. \text{ Thus } c^3 = b.$$

If you use a fine graph paper, it is easy to get three-digit accuracy in this construction.

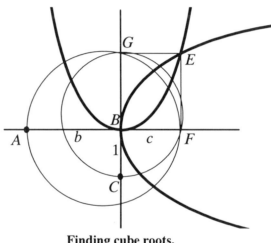

Figure 14.7   **Finding cube roots.**

The Greeks did a thorough study of conic sections and their properties which culminated in Apollonius's book *Conics* which appeared in 200 BC. You can read this book in English translation; see [A: Apollonius].

a. *Use the above geometric methods with a fine graph paper to find the cube root of 10.*

To find roots of cubic equations we shall also need to know *the (rectangular) hyperbola with vertex B and parameter AB.* This is the curve H, such that if E is on H and $ACED$ is the determined rectangle (see Figure 14.8), then $(EC)^2 = BC{\cdot}AC$.

The point E can be constructed using the construction from Chapter 13. Let F be the bisector of AB. Then the circle with center F and radius FC will intersect at D the line perpendicular to AB at A. From the drawing it is clear how these circles also construct the other branch of the hyperbola (with vertex A.)

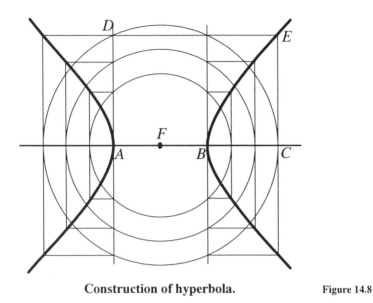

Construction of hyperbola.   Figure 14.8

b. *Use the above method with graph paper to construct the graph of the hyperbola with parameter 5. What is an algebraic equation that represents this hyperbola?*

Notice how these descriptions and constructions of the parabola and hyperbola look very much like they were done in Cartesian coordinates. *The ancestral forms of Cartesian coordinates and analytic geometry are evident here.* Also they are evident in the solutions of cubic equations in the next section. The ideas of Cartesian coordinates did not appear to Descartes out of nowhere. The underlying concepts were developing in Greek and Muslim mathematics. One of the apparent reasons that full development did not occur until Descartes is that, as we have seen, negative numbers were not accepted. The full use of negative numbers is essential for the realization of Cartesian coordinates.

PROBLEM 40. *Roots of Cubic Equations*

In his *Al-jabr wa'l muqabalah,* Omar Khayyam also gave geometric solution to cubic equations. You will see that his methods are sufficient to find geometrically all real (positive or negative) roots of cubic equations; however; in his first chapter Khayyam says (see [**A:** Khayyam (1931), page 49]):

> When, however, the object of the problem is an absolute number, neither we, nor any of those who are concerned with algebra, have been able to prove this equation — perhaps others who follow us will be able to fill the gap — except when it contains only the three first degrees, namely, the number, the thing and the square.

By "absolute number," Khayyam is referring to, what we call, algebraic solutions as opposed to geometric ones. This quotation suggests, contrary to what many historical accounts say, that Khayyam expected that algebraic solutions would be found.

Khayyam found 19 types of cubic equations (when expressed with only positive coefficients). (See [**A:** Khayyam (1931), page 51].) Of these 19, 5 reduce to quadratic equations (e.g., $x^3 + ax = bx$ reduces to $x^2 + ax = b$). The remaining 14 types Khayyam solves by using conic sections. His methods find all the positive roots of each type although he failed to mention some of the roots in a few cases; and, of course, he ignores the negative roots. Instead of going through his 14 types, I will show how a simple reduction will reduce all the types to only 3 types in addition to types already solved such as, $x^3 = b$. I will then give Khayyam's solutions to these three types.

In the cubic $y^3 + py^2 + gy + r = 0$ (where, here, $p, g, r,$ are positive, negative, or zero), set $y = x - (p/3)$. Try it! The resulting equation in x will have the form $x^3 + sx + t = 0$, (where, here, s and t are positive, negative, or zero). If we rearrange this equation so all the coefficients are positive then we get four types that have not been previously solved:

$$(1)\ x^3 + ax = b,\ (2)\ x^3 + b = ax,$$

$$(3)\ x^3 = ax + b,\ \text{and}\ (4)\ x^3 + ax + b = 0,$$

where a and b are positive, in addition to types previously solved.

a. *Show that in order to find all the roots of all cubic equations we need only have a method that finds the roots of Types 1, 2, and 3.*

Khayyam's solution for Type 1: $x^3 + ax = b$.

A cube and sides are equal to a number. Let the line *AB* [*see Figure 14.9*] be the side of a square equal to the given number of roots [*that is, $(AB)^2 = a$, the coefficient*]. Construct a solid whose base is equal to the square on *AB*, equal in volume to the given number [*b*]. The construction has been shown previously. Let *BC* be the height of the solid. [*I.e., $BC \cdot (AB)^2 = b$.*] Let *BC* be perpendicular to *AB* ... Construct a parabola whose vertex is the point *B* ... and parameter *AB*. Then the position of the conic *HBD* will be tangent to *BC*. Describe on *BC* a semicircle. It necessarily intersects the conic. Let the point of intersection be *D*; drop from *D*, whose position is known, two perpendiculars *DZ* and *DE* on *BZ* and *BC*. Both the position and magnitude of these lines are known.

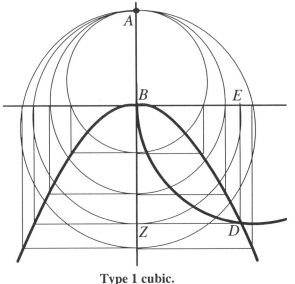

Type 1 cubic.   Figure 14.9

The root is *EB*. Khayyam's proof (using a more modern, compact notation) is: From the properties of the parabola (Problem 39) and circle (Chapter 13) we have

$$(DZ)^2 = (EB)^2 = BZ \cdot AB \text{ and } (ED)^2 = (BZ)^2 = EC \cdot EB ,$$

thus

$$EB \cdot (BZ)^2 = (EB)^2 \cdot EC = BZ \cdot AB \cdot EC$$

and therefore

$$AB \cdot EC = EB \cdot BZ$$

and

$$(EB)^3 = EB \cdot (BZ \cdot AB) = (AB \cdot EC) \cdot AB = (AB)^2 \cdot EC.$$

So

$$(EB)^3 + a(EB) = (AB)^2 \cdot EC + (AB)^2 \cdot (EB) = (AB)^2 \cdot CB = b.$$

Thus EB is a root of $x^3 + ax = b$. Since $x^2 + ax$ increases as x increases, there can be only this one root.

Khayyam's solutions for Types 2 and 3:

$$x^3 + b = ax \text{ and } x^3 = ax + b.$$

Khayyam treated these equations separately but by allowing negative horizontal lengths we can combine his two solutions into one solution of $x^3 \pm b = ax$. Let AB be perpendicular to BC and as before let $(AB)^2 = a$ and $(AB)^2 \cdot BC = b$. Place BC to the left if the sign in front of b is negative (Type 3) and place BC to the right is the sign in front of b is positive (Type 2). Construct a parabola with vertex B and parameter AB. Construct both branches of the hyperbola with vertices B and C and parameter BC.

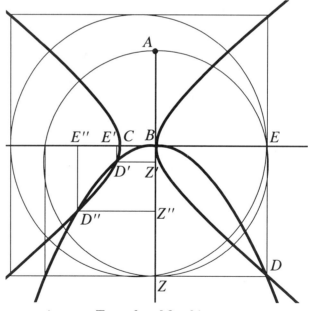

Figure 14.10   **Types 2 and 3 cubics.**

Each intersection of the hyperbola and the parabola (except for B) gives a root of the cubic. Suppose they meet at D. Then drop perpendiculars DE and DZ. The root is BE (negative if to the left and positive if to the right). Again, if you use fine graph paper, it is possible to get three digit accuracy here. I leave it for you, the reader, to provide the proof which is very similar to Type 1.

b. *Verify that Khayyam's method described above works for Types 2 and 3. Can you see from your verification why the extraneous root given by B appears?*

c. *Use Khayyam's method to find all solutions to the cubic*

$$x^3 = 15x + 4.$$

Use fine graph paper and try for three-place accuracy.

PROBLEM 41. *Algebraic Solution of Cubics*

A little more history: Most historical accounts assert correctly that Khayyam did not find the negative roots of cubics. However, they are misleading in that they all fail to mention that his methods are fully sufficient to find the negative roots as we have seen above. This is in contrast to the common assertion (see, for example, [**O:** Davis & Hersch]) that Girolamo Cardano (sixteenth-century Italian) was the first to publish the general solution of cubic equations when in fact, as we shall see, he himself admitted that his methods are insufficient to find the real roots of many cubics.

Cardano published his algebraic solutions in his book, *Artis Magnae* (The Great Art) which was published in 1545. For a readable English translation and historical summary, see [**A:** Cardano]. Cardano used only positive coefficients and thus divided the cubic equations into the same 13 types (excluding $x^3 = c$ and equations reducible to quadratics) used earlier by Khayyam. Cardano also used geometry to prove his solutions for each type. As we did above, we can make a substitution to reduce these to the same types as above:

$$(1)\ x^3 + ax = b,\ (2)\ x^3 + b = ax,$$

$$(3)\ x^3 = ax + b,\ \text{and}\ (4)\ x^3 + ax + b = 0.$$

If we allow ourselves the convenience of using negative numbers and lengths, then we can reduce these to one type: $x^3 + ax + b = 0$, where now we allow a and b to be either negative or positive.

The main "trick" that Cardano used was to assume that there is a solution of $x^3 + ax + b = 0$ of the form $x = t^{1/3} + u^{1/3}$. Plugging this into the cubic we get

$$(t^{1/3} + u^{1/3})^3 + a(t^{1/3} + u^{1/3}) + b = 0.$$

If you expand and simplify this, you get to

$$t + u + b + (3t^{1/3}u^{1/3} + a)(t^{1/3} + u^{1/3}) = 0.$$

(Cardano did this expansion and simplification geometrically by imagining a cube with sides $t^{1/3} + u^{1/3}$.) Thus $x = t^{1/3} + u^{1/3}$ is a root if

$$t + u = -b \quad \text{and} \quad t^{1/3}\,u^{1/3} = -(a/3).$$

Solving, we find that t and u are the roots of the quadratic equation

$$z^2 + bz - (a/3)^3 = 0$$

which Cardano solved geometrically (and so can you, Problem 38) to get

$$t = -b/2 + \sqrt{(b/2)^2 + (a/3)^3} \quad \text{and} \quad u = -b/2 - \sqrt{(b/2)^2 + (a/3)^3}\,.$$

Thus the cubic has roots

$$x = t^{1/3} + u^{1/3}$$

$$= \{-b/2 + \sqrt{(b/2)^2 + (a/3)^3}\,\}^{1/3} + \{-b/2 - \sqrt{(b/2)^2 + (a/3)^3}\,\}^{1/3}.$$

This is Cardano's cubic formula. But, a strange thing happened. Cardano noticed that the cubic $x^3 = 15x + 4$ has a positive real root 4 but, for this equation, $a = -15$ and $b = -4$, and if we put these values into his cubic formula, we get that the roots of $x^3 = 15x + 4$ are

$$x = \{2 + \sqrt{-121}\,\}^{1/3} + \{2 - \sqrt{-121}\,\}^{1/3}.$$

But these are complex numbers even though you have shown in Problem 40 that all three roots are real. How can this expression yield 4?

In Cardano's time there was no theory of complex numbers and so he reasonably concluded that his method would not work for this equation, even though he did investigate expressions such as $\sqrt{-121}$. Cardano writes ([A: Cardano, page 103]):

> When the cube of one-third the coefficient of x is greater than the square of one-half the constant of the equation ... then the solution of this can be found by the aliza problem which is discussed in the book of geometrical problems.

It is not clear what book he is referring to, but the "aliza problem" presumably refers to al'Hazen, an Arab, who lived around 1000 AD and whose works were known in Europe in Cardano's time. Al'Hazen had used intersecting conics to solve specific cubic equations and the problem of describing the image seen in a spherical mirror — this later problem is in some books called "Alhazen's problem."

In addition, we know today that each complex number has three cube roots and so the formula

$$x = \{2 + \sqrt{-121}\,\}^{1/3} + \{2 - \sqrt{-121}\,\}^{1/3}$$

is ambiguous. In fact, some choices for the two cube roots give roots of the cubic and some do not. (Experiment with $x^3 = 15x + 4$.) Faced with Cardano's Formula and equations like $x^3 = 15x + 4$, Cardano and other mathematicians of the time started exploring the possible meanings of these complex numbers and thus started the theory of complex numbers.

a. *Solve the cubic* $x^3 = 15x + 4$ *using Cardano's Formula and your knowledge of complex numbers.*

Remember that on the previous page we showed that $x = t^{1/3} + u^{1/3}$ is a root of the equation if $t + u = -b$ and $t^{1/3}u^{1/3} = -(a/3)$.

b. *Solve* $x^3 = 15x + 4$ *by dividing through by* $x - 4$ *and then solving the resulting quadratic.*

c. *Compare you answers and methods of solution from Problems* 40**c**, 41**a**, *and* 41**b**.

So What Does This All Point To?

So what does the experience of the last two chapters point to? It points to different things for each of us. I conclude that it is worthwhile paying attention to the meaning in mathematics. Often in our haste to get to the modern, powerful analytic tools we ignore and trod upon the meanings and images that are there. Sometimes it is hard even to get a glimpse that some meaning is missing. One way to get this glimpse and find meaning is to listen to and follow questions of "What does it mean?" that come up in ourselves, in our friends, and in our students. We must listen creatively because we and others often do not know how to express precisely what is bothering us.

Another way to find meaning is to read the mathematics of old and keep asking, "Why did they do that?" or "Why didn't they do this?" Why did the early algebraists (up until at least 1600 and much later, I think) insist on geometric proofs? I have suggested some reasons above. Today, we normally pass over geometric proofs in favor of analytic ones based on the 150-year-old notion of Cauchy sequences and the Axiom of Completeness. However, for most students and, I think, most mathematicians, our intuitive understanding of the real numbers is based on the geometric real line. As an example, think about multiplication: What does $a \times b$ mean? Compare the geometric images of $a \times b$ with the multiplication of two infinite, nonrepeating, decimal fractions. What is $\sqrt{2} \times \pi$?

There is another reason why a geometric solution may be more meaningful. Sometimes we want a geometric result instead of a numerical one. As an example, I shall describe an experience that I had while a friend and I were building a small house using wood. The roof of the house consisted of 12 isosceles triangles which together formed a 12-sided cone (or pyramid). It was necessary for us to determine the angle between two adjacent triangles in the roof so we could appropriately cut the log rafters. I immediately started to calculate the angle using (numerical) trigonometry and algebra. But then I ran into a problem. I had only a slide rule with three-place accuracy for finding square roots and values of trigonometric functions. At one point in the calculation I had to subtract two numbers that differed only in the third place (e.g., 5.68 - 5.65); thus my result had little accuracy. As I started to figure out a different computational procedure that would avoid the subtraction, I suddenly realized — *I didn't want a number, I wanted a physical angle.* In fact, a numerical angle would be essentially useless — imagine taking two rough boards and putting them at a given numerical angle apart using only an ordinary protractor! What I needed was the physical angle, full-size. So I constructed the angle on the floor of the house using a rope as a compass. This geometric solution had the following advantages over a numerical solution:

♦   The geometric solution resulted in the desired physical angle, while the numerical solution resulted in a number.

♦   The geometric solution was quicker than the numerical solution.

♦   The geometric solution was immediately understood and trusted by my friend (and fellow builder), who had almost no mathematical training, while the numerical solution was beyond my friend's understanding because it involved trigonometry (such as the "Law of Cosines").

♦   And, since the construction was done full-size, the solution automatically had the degree of accuracy appropriate for the application.

Meaning is important in mathematics and geometry is an important source of that meaning.

Chapter 15
Projections of a Sphere
onto a Plane

> Geography is a representation in picture of the whole known world together with the phenomena which are contained therein.
>
> ... The task of Geography is to survey the whole in its proportions, as one would the entire head. For as in an entire painting we must first put in the larger features, and afterward those detailed features which portraits and pictures may require, giving them proportion in relation to one another so that their correct measure apart can be seen by examining them, to note whether they form the whole or a part of the picture. ... Geography looks at the position rather than the quality, noting the relation of distances everywhere, ...
>
> It is the great and the exquisite accomplishment of mathematics to show all these things to the human intelligence ...
>
> — Ptolemy, *Geographia*, Book One, Chapter I

A major problem for map makers (cartographers) since Ptolemy and before is how to represent accurately a portion of the surface of a sphere on the plane. It is the same problem we have been having when making drawings to accompany our discussions of the geometry of the sphere. We shall use the terminology used by cartographers and differential geometers to call any one-to-one function from a portion of a sphere onto a portion of a plane a ***chart***. As Ptolemy states in the quote above, we would like to represent the sphere on the plane so that proportions (and thus angles) are preserved and the relative distances are accurate. It is impossible to make a chart without some distortions. *Which results that you have studied so far show that there must be distortions when attempting to represent a portion of a sphere on the plane?*

Nevertheless, there are projections from a portion of a sphere to the plane which take geodesics to straight lines, that is, which preserve the shape of straight lines. There are other projections which preserve all areas. There are still other projections which preserve the measure of all angles. In this chapter, we will study these types of projections.

PROBLEM 42. *Gnomic Projection*

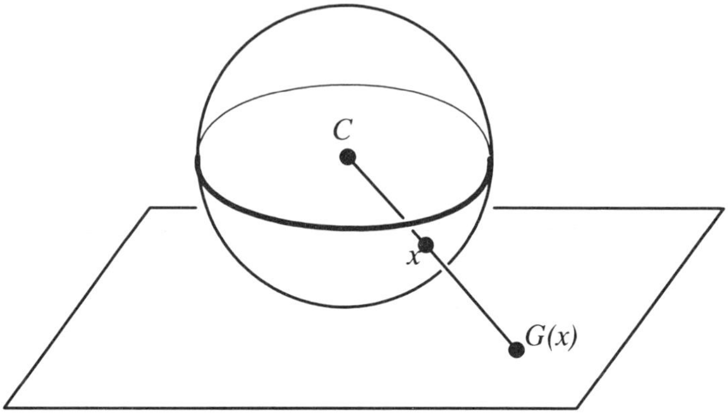

Figure 15.1   **Gnomic projection.**

Imagine a sphere resting on a horizontal plane. A *gnomic projection* is obtained by projecting from the center of a sphere onto the plane. Note that only the lower open hemisphere is projected onto the plane; i.e., if x is a point in the lower open hemisphere, then its gnomic projection is the point, $G(x)$, where the ray from the center through x intersects the plane.

*Show that a gnomic projection takes the portions of great circles in the lower hemisphere onto straight lines in the plane. (Because of this, a gnomic projection is said to be a **geodesic mapping**.)*

Gnomic projection is often used to make navigational charts for airplanes and ships. Why would this be appropriate?

Hint: Start with our extrinsic definition of great circle.

PROBLEM 43. *Cylindrical Projection*

Imagine a sphere of radius R, but this time center it in a vertical cylinder of radius R and height $2R$. The ***cylindrical projection*** is obtained by projecting from the axis of the cylinder which is also a diameter of the sphere; i.e., if x is a point (not the North or South Poles) on the sphere and $O(x)$ is the point on the axis at the same height as x, then x is projected onto the intersection of the cylinder with the ray from $O(x)$ to x.

Show that cylindrical projection preserves areas.

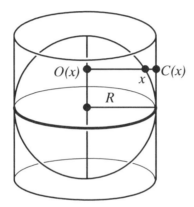

Cylindrical projection.   Figure 15.2

Geometric Approach: Look at an infinitesimal piece of area on the sphere bounded by longitudes and latitudes. Check that when it is projected onto the cylinder that the horizontal dimension becomes longer but the vertical dimension becomes shorter. Do these compensate for each other?

Analytic Approach: Find a function f from a rectangle in the (z, θ)-plane onto the sphere and a function g from the same rectangle onto the cylinder such that $C(f(z, \theta)) = g(z, \theta)$. Then use the techniques of finding surface area from vector analysis. (For two vectors A, B, the magnitude of the cross product $|A \times B|$ is the area of the parallelogram spanned by A and B. An element of surface area on the sphere can be represented by $|f_z \times f_\theta| \, dz \, d\theta$, the cross product of the partial derivatives.)

We can easily flatten the cylinder onto a plane and find its area to be $4\pi R^2$. We thus conclude:

The area of a sphere of radius R is $4\pi R^2$.

PROBLEM *44. Stereographic Projection*

Imagine the same sphere and plane, only this time project from the uppermost point (North Pole) of the sphere onto the plane. This is called *stereographic projection.*

*Show that stereographic projection preserves the sizes of angles. (Such mappings are called **conformal mappings**.)*

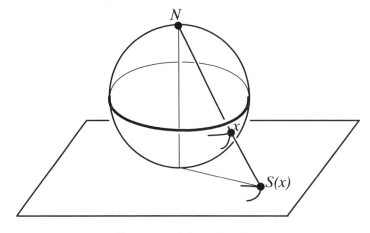

Figure 15.3   **Stereographic projection.**

Suggestions

There are several approaches for exploring this problem. Using a purely geometric approach requires visualization but only very basic geometry. An analytic approach requires knowledge of the differential of a function from $\mathbb{R}^2$ into $\mathbb{R}^3$.

Geometric Approach: An angle at a point x on the sphere is determined by two great circles intersecting at x. Look at the two planes that are determined by the North Pole and vectors tangent to the great circles at x. Notice that the intersection of these two planes with the horizontal image plane determines the image of the angle. Since the 3-dimensional figure is difficult for many of us to imagine in full detail, you may find it helpful to consider what is contained in various 2-dimensional planes. In particular, consider the plane determined by x and the North and South Poles, the plane tangent to the sphere at x, and the planes tangent to the sphere at the North and South Poles. Determine the relationships among these planes.

Analytic Approach: Introduce a coordinate system and find a formula for the function f from the plane to the sphere which is the inverse of S. Use the differential of f to examine the effect of f on angles. You will need to use the dot (inner) product and the fact that the differential of f is a linear transformation from the (tangent) vectors at $S(x)$ to the tangent vectors at x.

Chapter 16
Duality and Trigonometry

With a view to obtaining a table ready for immediate use, we shall next set out the lengths of these [chords in a circle], dividing the perimeter into 360 segments and by the side of the arcs placing the chords subtending them for every increase of half of a degree, that is, stating how many parts they are of the diameter, which it is convenient for the numerical calculations to divide into 120 segments.

—Ptolemy, *Syntaxis,* in [A: Thomas]

PROBLEM *45. Circumference of a Circle*

Find a simple formula for the circumference of a circle on a sphere in terms of its intrinsic radius and make the formula as intrinsic as possible.

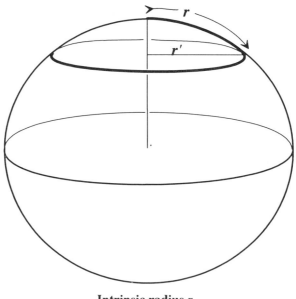

Figure 16.1　　　　　　**Intrinsic radius *r*.**

146

In Figure 16.1, rotating the segment of length r' (***the extrinsic radius***) through a whole revolution produces the same circumference as rotating r, which is an arc of the great circle as well as ***the intrinsic radius*** of the circle on the sphere.

We suggest that you make an extrinsic drawing (similar to Figure 16.1) of the circle, its intrinsic radius, its extrinsic radius, and the center of the sphere. You may well find it convenient to use trigonometric functions to express your answer. Even though the derivation of the formula this way will be extrinsic, it is possible, in the end, to express the circumference only in terms of intrinsic quantities. Thus, also think of the problem:

> *How could our 2-dimensional bug derive this formula?*

By looking at very small circles, the bug could certainly find uses for the trigonometric functions that they give rise to. Then the bug could discover that the geodesics are actually (intrinsic) circles, but circles which do not have the same trigonometric properties as very small circles. And then what? Use your experience from Chapter 11.

Note that the existence of trigonometric functions for right triangles follows from the properties of similar triangles that were proved in Problem 37.

PROBLEM 46. Law of Cosines

> *If we know two sides and the included angle of a (small) triangle, then according to SAS the third side is determined. If we know the lengths of the two sides and the measure of the included angle, how can we find the length of the third side? Find this length for triangles on both the plane and the sphere.*

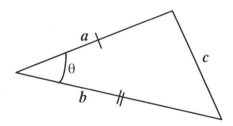

Law of Cosines.   **Figure 16.2**

You have learned in school (but perhaps forgotten) the *Law of Co-sines* on the plane: $c^2 = a^2 + b^2 - 2ab \cos(\theta)$. For a geometric proof of this "law," look at the pictures in Figure 16.3. These pictures show the squares as rigid with hinges at all the points marked ⟞⟝. Note that in the middle picture θ is greater than $\pi/2$. You must draw a different picture for θ less than $\pi/2$. Prove the Law of Cosines on the plane using the pictures in Figure 16.3, or in any other way you wish.

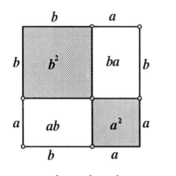

$$(a + b)^2 = a^2 + b^2 + 2ab$$

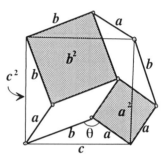

$$c^2 = a^2 + b^2 + 2ab(-\cos(\theta))$$

$$c^2 = a^2 + b^2$$

Figure 16.3   **Three related geometric proofs.**[†]

On the sphere there are various versions of the Law of Cosines that you can find. One approach that will work is to project the triangle by a gnomic projection onto the plane tangent to the sphere at the vertex of the given angle. This projection will preserve the size of the given angle (*Why?*) and, even though it will not preserve the lengths of the sides of the triangle, you can determine what effect it does have on these lengths.

[†] I first saw these pictures in the marvelous book [**G**: Valens].

Now apply the planar Law of Cosines to this projected triangle. It is very helpful to draw a 3-D picture of this projection.

It is often convenient to measure lengths of great circle arcs on the sphere in terms of the radian measure of the angle which the arc subtends at the pole of the great circle. We call this measure the ***radian measure of the arc***. For example, the radian measure of 1/4 great circle would be $\pi/2$ and the radian measure of half a great circle would be π. In Figure 16.4, the segment a is subtended by the angle α at the pole and by the same angle α at the center of the sphere. (Do you see why these angles are congruent? If not, imagine looking straight down on the sphere from above the pole.) The radian measure of a is the radian measure of α.

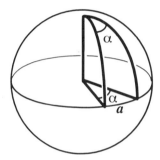

Radian measure of lengths.   **Figure 16.4**

Pause, explore, and write about this problem before you go to the next page.

If we measure lengths in radians, then one possible formula for the spherical triangle in Problem 46 is:

$$\cos c \; = \; \cos a \cos b \; + \; \sin a \sin b \cos \theta.$$

(This is not the only such formula.)

For a right triangle ($\theta = \pi/2$) the above formula becomes

$$\cos c \; = \; \cos a \cos b,$$

which can be considered as the spherical equivalent of the Pythagorean Theorem.

Closely related to the Law of Cosines is the Law of Sines.

PROBLEM 47. Law of Sines

If △ABC is a triangle on the plane with sides, a, b, c, and corresponding opposite angles, α, β, γ, then

$$a/\sin \alpha = b/\sin \beta = c/\sin \gamma.$$

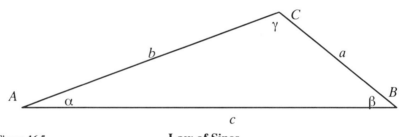

Figure 16.5   **Law of Sines.**

What is an analogous property on the sphere?

The standard proof for the Law of Sines is to drop a perpendicular from the vertex C to the side c and then to express the length of this perpendicular as both ($b \sin \alpha$) and ($a \sin \beta$). From this the result easily follows.

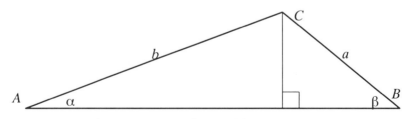

Figure 16.6   **Standard proof of Law of Sines on plane.**

Thus, on the plane the Law of Sines follows from an expression for the sine of an angle in a right triangle. For triangles on the sphere we can find a very similar result. If $\triangle ACD$ is a triangle on the sphere with the angle at D being a right angle, then use gnomic projection to project $\triangle ACD$ onto the plane which is tangent to the sphere at A. Since the plane is tangent to the sphere at A, the size of the angle α is preserved under the projection. In general, angles on the sphere not at A will not be projected to angles of the same size, but in this case the right angle at D will be projected to a right angle. (*Be sure you see why this is the case. A good drawing and the use of symmetry and similar triangles will help.*) Now express the sine of α in terms of the sides of this projected triangle.

Pause, explore, and write about this problem before you go to the next page.

When we measure the sides in radians, on the sphere the Law of Sines becomes:

$$(\sin a)/\sin \alpha = (\sin b)/\sin \beta = (\sin c)/\sin \gamma.$$

For a right triangle this becomes:

$$\sin \alpha = (\sin a)/(\sin c).$$

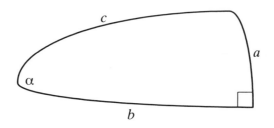

Figure 16.7   **Law of Sines for right triangles on a sphere.**

We can now ask: Is it possible to find expressions corresponding to the other triangle congruence theorems which we proved in Chapters 5, 7, and 9. Let us see how the concept of duality will help us.

Duality on a Sphere

When we were looking at SAS and ASA, we noticed a certain duality between points and lines (geodesics). SAS was true on the plane or open hemisphere because *two points determine a unique line segment* and ASA was true on the plane or open hemisphere because *two (intersecting) lines determine a unique point*. In this section we will make this notion of duality broader and deeper and look at it in such a way that it applies to both the plane and the sphere.

On the whole sphere two distinct points determine a unique straight line (great circle) *unless the points are antipodal.* In addition, two distinct great circles determine a unique *pair of antipodal points.* Also, a circle on the sphere has two centers *which are antipodal.* Remember also that in most of the triangle congruence theorems, we had *trouble with triangles which contained antipodal points.* So our first step is to consider not points on the sphere but rather ***point-pairs***, pairs of antipodal points. With this definition in mind, check the following:

♦   *Two distinct point-pairs determine a unique great circle (geodesic).*

♦   *Two distinct great circles determine a unique point-pair.*

◆   *The center of a circle is a single point-pair.*

◆   *SAS, ASA, SSS, AAA are true for all triangles not containing any point-pairs.*

Now, we can make the duality more definite:

◆   *The dual[†] of a great circle is its* **poles** (the point-pair that is the intrinsic center of the great circle).

◆   *The dual of a point-pair is its* **equator** (the great circle whose center is the point-pair).

Notice:

If the point-pair **P** *is on the great circle* **l**, *then the dual of* **l** *is on the dual of* **P**.

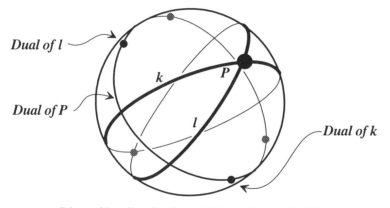

P is on *l* implies the dual of *l* is on the dual of P.   **Figure 16.8**

If *k* is another great circle through *P*, then notice that the dual of *k* is also on the dual of *P*. Since an angle can be viewed as a collection of lines (great circles) emanating from a point, the dual of this angle is a collection of point-pairs lying on the dual of the angle's vertex. And, vice versa, the dual of the points on a segment of a great circle are great circles emanating from a point-pair which is the dual of the original great circle. *Before going on be sure to understand this relationship between an angle and its dual.* Draw pictures. Make models.

[†] Some books use the term "*polar*" in place of "*dual*."

PROBLEM 48. *The Dual of a Small Triangle*

The dual of the small triangle $\triangle ABC$ *is the small triangle* $\triangle A*B*C*$, *where* $A*$ *is that pole of the great circle of* BC *which is on the same side of* BC *as the vertex* A, *similarly for* $B*$ *and* $C*$. *Figure 16.9.*

Find the relationship between the sizes of the angles and sides of a triangle and the corresponding sides and angles of its dual.

Is there a triangle which is its own dual?

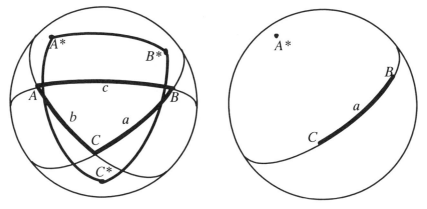

Figure 16.9   **The dual of a small triangle.**

Using duality and your formulas from Problems 46 and 47, you can now work on the next problem.

PROBLEM 49. *Trigonometry on Spherical Triangles*

For each of ASA, RLH, SSS, AAA, if you know the measures of the given sides and angles, how can you find the measures of the sides and angles that are not given?

Duality on the Projective Plane

The gnomic projection, G, (Problem 42) allows us to transfer the above duality on the sphere to a duality on the plane. If P is a point on the plane, then there is a point Q on the sphere such that $G(Q) = P$. The dual of Q is a great circle g on the sphere. If Q is not the South Pole, then half of g is in the Southern Hemisphere and its projection onto the plane, $G(g)$, is a line which we can call the *dual of P*. This defines a dual for every point on the plane except for the point where the South Pole of the sphere rests. See Figure 16.10. It is convenient to call this point the origin, O, of the plane.

Note that O is the image of S, the South Pole, and that the dual of S is the equator which is projected by G to infinity on the plane. Thus we define the dual of O to be the **line at infinity**. If l is any line in the plane, then it is the image of a great circle on the sphere which intersects the equator in a point-pair. The image of this point-pair is considered to be a *single* point at infinity at the "end" of the line l, the *same* point at both ends. The plane with the line at infinity attached is called the **projective plane**.

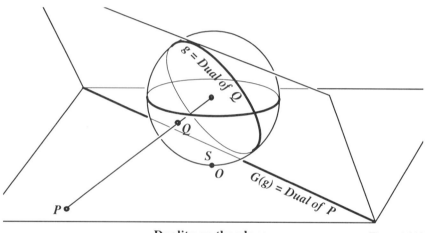

Duality on the plane.   Figure 16.10

PROBLEM 50. *Properties on the Projective Plane*

Check that the following hold on the projective plane:

a. Two points determine a unique line.

b. Two parallel lines share the same point at infinity.

c. Two lines determine a unique point.

d. If a point is on a line, then the dual of the line is a point which is on the dual of the original point.

e. If C is the circle with center at the origin and with radius the same as the radius of the sphere, then the dual of a point on C is a line tangent to C.

Perspective Drawings and Vision

Look at the following perspective drawing:

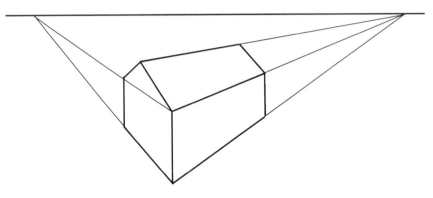

Figure 16.11   **Perspective drawing.**

Present day theories of projective geometry got their start from Euclid's *Optics* [**A**: Euclid] and later from the theories of perspective developed during the Renaissance by artists who studied the geometry inherent in perspective drawings. The "vanishing point" where the lines which are parallel on the house intersect on the horizon of the drawing is an image on the drawing of the point at infinity on these parallel lines.

One way to visualize this is to imagine yourself at the center of a transparent sphere looking out at the world. If you look at two parallel straight lines and trace these lines on the sphere, you will be tracing segments of two great circles. If you followed this tracing indefinitely to the ends of the straight lines, then you would have a tracing of two half great circles on the sphere intersecting in their endpoints. These endpoints of the semicircles are the images of the point at infinity that is the common intersection of the two parallel lines. If you now use a gnomic projection to project this onto a plane (e.g., the artist's canvas), then you will obtain two straight line segments intersecting at one of their endpoints as in the drawing above.

Chapter 17
Isometries and Patterns

All shapelessness whose kind admits of pattern and form, as long as it remains outside of Reason and Idea, is ugly by that very isolation from the Divine Reason-Principle. And this is the Absolute Ugly: an ugly thing is something that has not been entirely mastered by pattern, that is by Reason, the Matter not yielding at all points and in all respects to Ideal-Form.

— Plotinus, *The Enneads*, I.6.2 [**A**: Plotinus]

In this chapter we will look at some group theory through its origins, that is, geometrically. First, we will introduce some definitions. We advise you to investigate them as concretely as possible.

Definitions and Terminology

An *isometry* is a one-to-one transformation from one geometric space onto itself which preserves all intrinsic properties such as length, angle measure, and area. Note that reflections, rotations, translations, and any compositions of them are isometries. In Figure 17.1, there are given two geometric figures, $\mathscr{F}$ and $\mathscr{G}$ which are congruent, but there is not a single reflection or rotation or translation which will take one onto the other. However, it is clear that there is some composition of translations, rotations, and reflections which will take $\mathscr{F}$ onto $\mathscr{G}$. In fact, the composition of a reflection through the line l and a translation along l will take $\mathscr{F}$ onto $\mathscr{G}$. This composition is called *a glide reflection along l.*

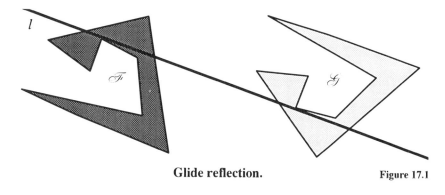

Glide reflection.   Figure 17.1

In Chapter 1 we talked about symmetries of the line. All of those symmetries can be seen as isometries of the plane except for similarity symmetry and 3-D rotation symmetry (through any angle not an integer multiple of 180°). Similarity symmetry changes lengths between points of the geometric figure and is thus not an isometry. Three-dimensional rotation symmetry is an isometry of 3-space, but it moves any plane off itself and thus cannot be an isometry of a plane (unless the angle of rotation is a multiple of 180°).

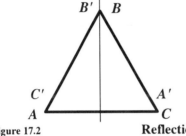

What is a symmetry of a geometric figure? A *symmetry* of a geometric figure is an isometry which takes the figure onto itself. For example, reflection though any median is a symmetry of an equilateral triangle.

Figure 17.2   **Reflection symmetry.**

Every isometry is a symmetry of some geometric figure. If we take a geometric figure, often called the *motif*, and apply to it the isometry and its inverse making additional copies of the motif over and over again, then we obtain another geometric figure for which the initial isometry is a symmetry. Let us look at an example:

If we start with the motif   and the isometry is translation

to the right through a distance *d*, then, using this isometry and its inverse (translation to the left through a distance *d*) we obtain the pattern:

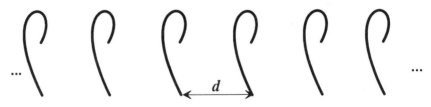

Figure 17.3   **Translation symmetry..**

If the isometry is clock-
wise rotation through
1/3 of a revolution
about the lower end-
point of the motif, then
we obtain the pattern:

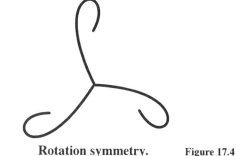

Rotation symmetry.   Figure 17.4

In the examples above we have generated two "patterns"; the upper
one has for a generator the translation, T_d. The lower one is generated by
a rotation, $S_{1/3}$. We generated two different patterns. But what is a
pattern?

A ***pattern*** is a figure <u>together</u> <u>with</u> <u>all</u> its <u>symmetries</u>. For example,
the upper figure above is a pattern with symmetries {Id (the identity),
and T_{nd}, where $n = \pm1, \pm2, \pm3, ...$}. The lower figure above is a pattern
with symmetries {Id, $S_{1/3}$, $S_{2/3}$}. An equilateral triangle with its six
symmetries { $R_A, R_B, R_C, S_{1/3}, S_{2/3}, $Id } is a pattern, where R_A is the reflection through the median from A.

We call the collection of all symmetries of a geometric figure its
symmetry group. If g, h are symmetries of a figure $\mathcal{F}$, then you can
easily see that:

> *The **composition** gh (first transform by h and then follow
> it by g) is also a symmetry. For example, for an equilateral
> triangle the composition of the reflection R_A with the
> reflection R_B is a rotation, $S_{2/3}$. In symbols, $R_B R_A = S_{2/3}$.*

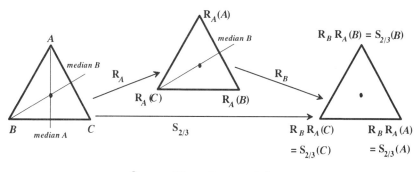

Composition of symmetries.   Figure 17.5

*The **identity** transformation,* Id *(the transformation that takes every point to itself) is a symmetry.*

*For every symmetry g of a figure there is another symmetry f such that gf and fg are the identity — in this case we call f the **inverse** of g. For example, the inverse of the rotation* $S_{2/3}$ *is the rotation* $S_{1/3}$ *and vice versa. In symbols,* $S_{1/3}S_{2/3} = S_{2/3}S_{1/3} = \text{Id}.$

We say that two patterns are the same (or *isomorphic*) if they have the same symmetries. However, this does not mean that if two patterns are isomorphic, then the corresponding geometric figures have to be congruent. For example, the pattern on the left in Figure 17.6 is isomorphic to the equilateral triangle whereas the patterns in the middle and on the right are not.[†]

Figure 17.6   **Non-isomorphic patterns.**

A *strip* (or linear, or frieze) *pattern* is a pattern which has a translation symmetry and all of whose symmetries are also symmetries of a given line. For example, the following is a strip pattern with symmetries: (1) reflection through *l*, and (2) translations through distances *nd*, $n = 0, \pm 1, \pm 2, \pm 3, \cdots$.

Figure 17.7   **A strip pattern.**

You are now able to start to study properties about patterns and isometries.

[†]If two patterns are isomorphic, then their symmetry groups are isomorphic as abstract groups. The converse is often, but not always, true. For example, the symmetry groups of the letter A and the letter S are both isomorphic as abstract groups to Z_2, but are not isomorphic as patterns.

PROBLEM 51. *Examples of Patterns*

> *Find as many (non-isomorphic) patterns as you can which have only finitely many symmetries.*

> *Find as many (non-isomorphic) strip patterns as you can.*

> *List the symmetries of each.*

Suggestions

Examples of different strip patterns and many finite patterns can be found on buildings everywhere: houses of worship, courthouses, and most older buildings. Also look at other decorations on plates or on wallpaper edging.

In order to determine if your list of strip patterns and finite patterns contains all possible strip and finite patterns, we need first to explore properties of isometries. Here is one very important property of isometries of a plane: If two isometries act on three non-colinear points of the plane in the same way, then they are the same isometry. Let us see an example of how this works before you see why this property holds. In the Figure 17.8, H_p represents the half-turn around p, R_m represents the reflection through line m, G_l represents a glide reflection along l, and

$$\mathscr{T}_1 = \{A, B, C\}, \quad \mathscr{T}_2 = \{A', B', C'\} \text{ and } \mathscr{T}_3 = \{A'', B'', C''\}.$$

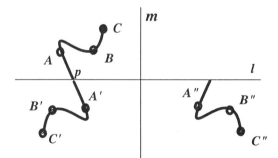

A glide equals a half-turn followed by a reflection.   Figure 17.8

We can see that $H_p(\mathscr{T}_1) = \mathscr{T}_2$, $R_m(\mathscr{T}_2) = \mathscr{T}_3$, and that $G_l(\mathscr{T}_1) = \mathscr{T}_3$. But then G_l and $R_m H_p$ perform the same action on the three points. If we use Problem 52, we can say that $G_l = R_m H_p$; that is, $G_l(X) = R_m H_p(X)$, for all points X on the plane. You can see now the usefulness of proving Problem 52.

PROBLEM 52. *Isometry Determined by Its Action on Three Points*

> *Prove the following: On the plane or sphere, if f and g are isometries and A, B, C are three non-collinear points, such that f(A) = g(A), f(B) = g(B), and f(C) = g(C), then f and g are the same isometry, that is, f(X) = g(X) for every point X.*

Suggestions

If you have trouble getting started with this problem, then take a specific example such as the two congruent figures, $\mathcal{F}$ and $\mathcal{G}$ at the start of this chapter and label three non-collinear points, A, B, C. Pick another point X and convince yourself as concretely as possible that any sequence of isometries that takes $\mathcal{F}$ onto $\mathcal{G}$ must always take X to the same location.

PROBLEM 53. *Classification of Isometries on Plane and Sphere*

> *Prove the following through geometric constructions:*
>
> *(a) Every isometry is the composition of one, two, or three reflections.*
>
> *(b) Every composition of two reflections is either a translation, a rotation, or the identity. How can you tell which one?*
>
> *(c) Every composition of three reflections is either a reflection or a glide reflection. How can you tell which one? A **glide reflection** is defined as a reflection through a line l composed with a translation along the same l.* —l

If (a) through (c) are true, then:

> *(d) Every isometry of the plane or sphere is either a reflection, a translation, a rotation, a glide reflection, or the identity.*

Suggestions

In order to prove Problem 53, you need to use Problem 52. Problem 52 allows you to be more concrete when thinking about Problem 53 because you only need to look at the effect of the isometries on any three non-collinear points that you pick. To get started on (a), cut a triangle out of an index card and use it to draw two congruent triangles in different orientations on a sheet of paper. For example:

Can you make triangles coincide with three reflections?   Figure 17.9

Now, can you move one triangle to the other by three (or less) reflections? You can use your cutout triangle for the intermediate steps.

It is also possible to solve this problem by looking at pairs of congruent triangles and directly showing that (d) is true.

You now have powerful tools to make a classification of discrete strip patterns on the plane and sphere and finite patterns on the plane. A strip pattern is *discrete* if every translation symmetry of the strip pattern is a multiple of some shortest translation.

PROBLEM 54. *Classification of Discrete Strip Patterns*

Prove there are only seven strip patterns on the plane which are discrete.

What are some non-discrete strip patterns?

What happens with strip patterns on a sphere?

Hint: Use Problem 53.

PROBLEM 55. *Classification of Finite Plane Patterns*

Any pattern on the plane with only finitely many symmetries has a center. That is, there is a point in the plane (not necessarily on the figure) such that every symmetry of the pattern leaves the point fixed. Is this true on the sphere?

Describe all the patterns on the plane with only finitely many symmetries.

This was first proved by Leonardo daVinci, and you can prove it too!

If you take a cube with its vertices on a sphere and project from the center of the sphere the edges of the cube onto the sphere, then the result is a pattern on the sphere with only finitely many symmetries. This pattern does not fit with the plane patterns you found in Problem 55 because this pattern has no center (on the sphere). See Problem 60 for more examples.

Geometric Meaning of Some Abstract Group Terminology

The collection of symmetries of a geometric figure with the operation of composition is an ***abstract group.*** We showed above how the usual axioms of a group are satisfied. For example, the following figure has symmetry group $\{ R_A, R_B, R_C, S_{1/3}, S_{2/3}, \text{Id} \}$ which is isomorphic as an abstract group to D_3 the third dihedral group.

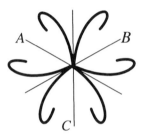

Figure 17.10   Isomorphic to D_3.

If a figure has a subfigure, then the symmetry group of the subfigure is a ***subgroup*** of the symmetry group of the original figure. For example:

   is a subfigure of

Figure 17.11   Subgroups.

and the symmetry group of the subfigure is $\{ S_{1/3}, S_{2/3}, \text{Id} \}$ which is iso-morphic as an abstract group to Z_{t_3} which is isomorphic to a subgroup of $\mathbf{D_3}$.

In $\mathbf{D_3}$ the two *cosets* of Z_{t_3}, $\text{Id}Z_{t_3}$ and $R_A Z_{t_3} = R_B Z_{t_3} = R_C Z_{t_3}$, cor-respond to the two copies of the subfigure:

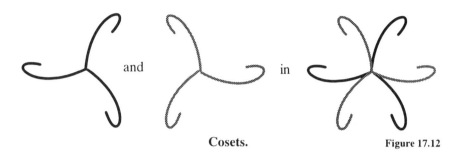

and   in

Cosets.   Figure 17.12

In general, if G is the symmetry group of a figure and H is a sub-group which is the symmetry group of a subfigure in the figure, then the *cosets* of H correspond to the several congruent copies of the subfigure that exist within the larger figure.

Chapter 18
Polyhedra

> The right way is next in order after the second dimension to take the
> third. This, I suppose, is the dimension of cubes and of everything
> that has depth.
>
> — Plato, *Republic*, VII.528b [**A**: Plato]

Definitions and Terminology

A *tetrahedron*, $\triangle ABCD$, in 3-space [in a 3-sphere][†] is determined
by any four points, A,B,C,D, called its **vertices**, such that all four points
do not lie on the same plane [great 2-sphere] and no three of the points lie
on the same line [great circle]. The *faces* of the tetrahedron are the four
[small] triangles $\triangle ABC$, $\triangle BCD$, $\triangle CDA$, $\triangle DAB$. The *edges* of the tetrahedron are the six line [great circle] segments AB, AC, AD, BC, BD, CD.
The *interior* of the tetrahedron is the [smallest] 3-dimensional region that
it bounds.

Tetrahedra are to three-dimensions as triangles are to two-
dimensions. Every polyhedron can be dissected into tetrahedra, but the
proofs are considerably more difficult than the ones from Problem 11,
and in the discussion to Problem 11 there is a polyhedron that is impossible to dissect into tetrahedra without adding extra vertices. There are numerous congruence theorems for tetrahedra, analogous to the congruence
theorems for triangles. All of the problems below apply to tetrahedra in
Euclidean 3-space or on a 3-sphere.

The *dihedral angle*, $\angle AB$, at the edge AB is the angle formed at
AB by $\triangle ABC$ and $\triangle ABD$. The dihedral angle is measured by intersecting it with a plane which is perpendicular to AB at a point between A
and B. The *solid angle* at A, $\angle A$, is that portion of the interior of the
tetrahedron "at" the vertex A. See Figure 18.1.

[†]In this chapter the text in the brackets applies to polyhedra on a 3-sphere.

166

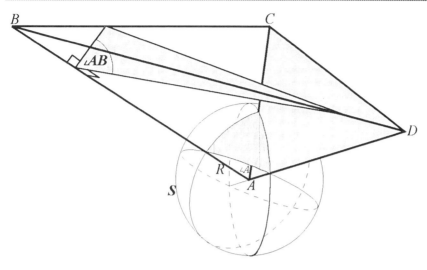

Dihedral and solid angles.   **Figure 18.1**

Problem 56. *Measure of a Solid Angle*

The measure of the solid angle is defined as the ratio,

$$m(\angle A) = [\lim_{R\to 0}] \text{ area}\{ \text{ (interior of } \triangle ABCD) \cap S \} / R^2 ,$$

where S is any small 2-sphere with center at A whose radius, R, is smaller than the distance from A to each of the other vertices.

Show that the measures of the solid and dihedral angles of a tetrahedron satisfy the following relationship:

$$m(\angle A) = m(\angle AB) + m(\angle AC) + m(\angle AD) - \pi.$$

Show that two solid angles with the same measure are not necessarily congruent.

Suggestions

Solid angles, whether in Euclidean 3-space or on a 3-sphere, are closely related to spherical triangles on a small sphere around the vertex. You can think of starting with a sphere, S, and creating a solid angle by extending three sticks out from the center of the sphere. If you connect the ends of these sticks, you will have a tetrahedron. The important thing to notice is how the sticks intersect the sphere. They will obviously intersect the sphere at three points, and you can draw in the great circle arcs connecting these points. Look at the planes in which the great circles lie.

In this problem you need to figure out the relationships between the angles of the spherical triangle and the dihedral angles.

The formula given above for the definition of the solid angle uses the intersection of the interior of the solid angle with any small sphere S. This intersection is the small triangle that you just drew, and the area of the intersection is the area of the triangle. Since the measure of a solid angle is defined in terms of an area, it is possible for two solid angles to have the same measure without being congruent — they can have the same area without having the same shape.

What you are asked to prove here is the relationship between the measure of the solid angle and the measures of the dihedral angles. Since they are closely related to spherical triangles on the small sphere, you can use everything you know about small triangles on a sphere.

PROBLEM 57. *Edges and Face Angles*

If △ABCD and △A'B'C'D' are two tetrahedra such that

$$\angle BAC \cong \angle B'A'C', \; \angle CAD \cong \angle C'A'D',$$

$$\angle BAD \cong \angle B'A'D',$$

$$CA \cong C'A', \; BA \cong B'A', \; DA \cong D'A'$$

then △ABCD ≅ △A'B'C'D'.

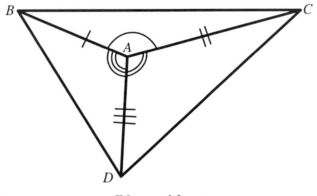

Figure 18.2   **Edges and faces.**

Part of your proof must be to show that the solid angles ∠A and ∠A' are congruent and not merely that they have the same measure.

Suggestions

If S is a small sphere with center at A and radius R, then

$$S \cap (\text{interior of } \Delta ABCD)$$

is a spherical triangle whose sides have lengths

$$R \times \angle BAC, \, R \times \angle CAD, \, R \times \angle BAD.$$

In the last problem, you saw how solid angles are related to spherical triangles. This problem asks you to prove the congruence of tetrahedra based on certain angle and length measurements. (Note that the angles shown above are not the dihedral angles of the tetrahedron.) So, since you can use spherical triangles to relate solid and dihedral angle measurements, why not use them to prove tetrahedra congruencies? Use the hint given to see what measurements of the spherical triangle are defined by measurements of the tetrahedron. Then see if the measurements given do in fact show congruence, and show why.

PROBLEM 58. *Edges and Dihedral Angles*

If

$$AB \cong A'B', \, \angle AB \cong \angle A'B', \, AC \cong A'C',$$

$$\angle AC \cong \angle A'C', \, AD \cong A'D', \, \angle AD \cong \angle A'D',$$

then

$$\Delta ABCD \cong \Delta A'B'C'D'.$$

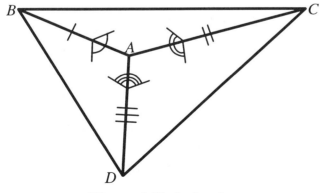

Edges and dihedral angles.   **Figure 18.3**

This is very similar to the previous problem, but uses different measurements — here we have the dihedral angles instead of the angles on the faces of tetrahedron. Look at this problem the same way you looked at the previous one — see how the measurements given relate to a spherical triangle, and then prove the congruence.

PROBLEM 59. *Other Congruence Theorems for Tetrahedra*

Make up your own congruence theorems! Find and prove at least two other sets of conditions that will imply congruence for tetrahedra, that is, make up and prove other theorems like those in Problems 57 and 58.

It is important that you make sure your conditions are sufficient to prove that the solid angles are congruent, not just that they have the same measure.

PROBLEM 60. *The Five Regular Polyhedra*

A *regular polygon* is a polygon lying in a plane or 2-sphere such that all of its edges are congruent and all of its angles are congruent. For example on the plane a regular quadrilateral is a square. On a 2-sphere a regular quadrilateral is constructed as follows:

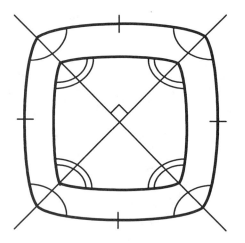

Figure 18.4   **Regular quadrilaterals.**

Note that half of a regular quadrilateral is a Khayyam quadrilateral. On a 2-sphere there are no similar polygons; for example, a regular quadrilateral (congruent sides and congruent angles) will have the same angles as another regular quadrilateral if and only if they have the same area. (Do you see why?)

A *polyhedron* in 3-space [or in a 3-sphere] is *regular* if all of its edges are congruent, all of its face angles are congruent, all of its dihedral angles are congruent, and all of its solid angles are congruent. The faces of a polyhedron are assumed to be polygons that lie on a plane [on a great 2-sphere].

> *Show that there are only five regular polyhedra. In Euclidean 3-space, to say "there are only five regular polyhedra" is to mean that any regular polyhedra is similar (same shape, but not necessarily the same size) to one of the five. It still makes sense on a 3-sphere to say that "there are only five regular polyhedra," but you need to make clear what you mean by this phrase.*

These polyhedra are often called "the Platonic Solids," and are described by Plato as "forms of bodies which excel in beauty;"[†] but there is considerable evidence that they were known well before Plato's time.[††] The regular polyhedra are the subject of Euclid's thirteenth (and last) book.

Suggestions

Your argument should be essentially the same whether you are considering a 3-sphere or 3-space. There are many widely different ways to do this problem. Here we suggest one approach. First note that the faces of a regular polyhedron must be regular polygons. Then focus on the vertices of regular polyhedra. Show that if the faces are regular quadrilaterals or regular pentagons, then there must be precisely three faces intersecting at each vertex.

Show that it is impossible for regular hexagons to intersect at a vertex to form the solid angle of a regular polyhedron. If the faces are regular (equilateral) triangles, then show that there are three possibilities at the vertices.

You may find it helpful to review the chapters on holonomy and area, isometry and patterns, 3-spheres in 4-space, and the first part of this chapter.

The five regular polyhedra are usually named the Tetrahedron, the Cube, the Octahedron, the Dodecahedron, and the Icosahedron. (See Figure 18.5.) There is a duality (related to but not exactly the same as the

---

[†] *Timaeus*, 53e [**A**: Plato].
[††] See T. L. Heath's discussion of the evidence in [**A**: Euclid, Vol. 3, pp. 438-9].

duality in Chapter 15 on duality and trigonometry) among regular polyhedra: If you pick the centers of the faces of a regular polyhedron, then
these points are the vertices of a regular polyhedron which is called the
dual of the original polyhedron. You can see that the cube is dual to the
octahedron (and vice versa), that the icosahedron is dual to the dodecahedron (and vice versa), and that the tetrahedron is dual to itself. See Figure
18.5

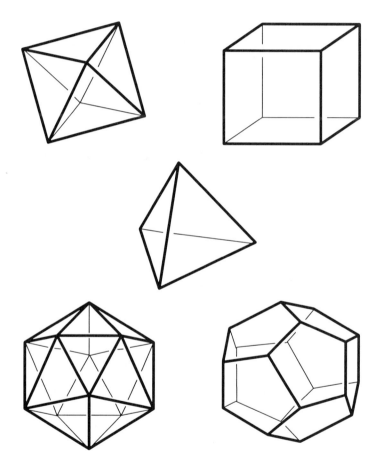

Figure 18.5   **The five Platonic solids.**

Appendix

A Geometric Introduction to Differential Geometry

As is well known, geometry presupposes the concept of space, as well as assuming the basic principles for constructions in space. It gives only nominal definitions of these things, while their essential specifications appear in the form of axioms. The relationship between these presuppositions is left in the dark; we do not see whether, or to what extent, any connection between them is necessary, or a priori whether any connection between them is even possible.

From Euclid to the most famous of the modern reformers of geometry, Legendre, this darkness has been dispelled neither by the mathematicians nor by the philosophers who have concerned themselves with it. This is undoubtedly because the general concept of multiply extended quantities, which includes spatial quantities, remains completely unexplored. I have therefore first set myself the task of constructing the concept of a multiply extended quantity from general notions of quantity. It will be shown that a multiply extended quantity is susceptible of various metric relations, so that Space constitutes only a special case of a triply extended quantity. ...

— Georg Riemann, *On the Hypotheses Which Lie at the Foundations of Geometry*, translated from the German in [C: Spivak, Vol. II, pp. 135]

The formalism of differential geometry is considered by many to be one of the most complicated and inaccessible of all the formal systems in mathematics. It is probably fair to say that most mathematicians do not feel comfortable with their understanding of differential geometry.

In this chapter we will give geometric descriptions that follow the spirit of this book of some basic notions and theorems in differential geometry. We shall leave to the reader who has studied differential geometry the task of showing how these geometric descriptions connect to whichever version of the analytic and formal descriptions was studied.

The Universe — Zooms

It is assumed, for now, that our discussions take place within a universe which is a usual *Euclidean space* wherein any two points, p and

q, determine a unique straight line segment which we shall denote $[p,q]$. If p is any point, then the collection of line segments with p as an end point are the vectors of a *vector space* with p as its origin. When the linear algebraic properties of the Euclidean space are being emphasized, it is usually called an *affine linear space*. For now the dimension of this universe can be left as indeterminant.

As is our normal experience when we look at this space, only a certain region of it is within our *field of view*. For simplicity we shall assume that all the fields of view are the shape of a round (spherical) ball and that we see all parts of this ball with equal clarity. (That is, we ignore what we usually call peripheral vision, which is a region at the edge of our field of view where we can see less detail than in the center of the field of view.) We express the *tolerance* of our vision as a percentage τ by specifying that two points in the field of view (f.o.v.) are *indistinguishable* (in the field of view) if their distance apart is less than $1/\tau$ times the radius of the f.o.v. When we focus in on a portion of space, we decrease the size of our f.o.v., but we shall assume that the tolerance (as a percentage of the radius) stays the same.

For now, what is important in this universe is that we can *zoom in on a point*, p, that is, we can decrease the size of our f.o.v. keeping the point p always in the center of the f.o.v. as happens, for example, with a zoom lens on a microscope or zooms in many computer graphics environments.

Smooth Curves

A *differentiable curve* is a geometric object which, for any clarity, if you zoom in on any point, p, becomes indistinguishable from a straight line which we call the *tangent line at p*, or T_p. We shall call it a *smooth curve* if the amount of zooming necessary is *uniform* in the sense that for any tolerance there is a length δ such that if we center a f.o.v. of radius δ at any point, p, on the curve then within that f.o.v. the curve is indistinguishable from the tangent line at p. Within a particular f.o.v. containing p, one may approximate T_p by the line which is determined by two points (distinguishable in the f.o.v.) lying on either side of p along the curve.

If a curve has the same tangent line at every point, then the curve is a *straight line*. The *osculating circle* C_p at p, if it exists, is a circle such that, in some f.o.v., the circle is indistinguishable from the curve along an interval of the curve which is not straight. The radius, r, of the osculating circle is called the *radius of curvature at p* and the vector which has magnitude (length) $1/r$ and direction the direction from p to the center of the osculating circle is called the *curvature* κ. We may approximate C_p by picking two (not collinear with p) points, p-h and p+h, on either side of p

at a distance h from p along the curve. Then the circle determined by $p-h$, p, $p+h$ approximates C_p. We may see in the following picture that the line through $p-h$ and p approximates the tangent line at $p-h/2$ and the line through p and $p+h$ approximates the tangent line at $p+h/2$. Let $\Delta\mathbf{T}$ be the change in the unit tangent vector from $p-h$ to $p+h$. Thus, by definition, the rate of change of the unit tangent vector is approximated by $\Delta\mathbf{T}/h = \alpha/h = 1/r = |\kappa|$.

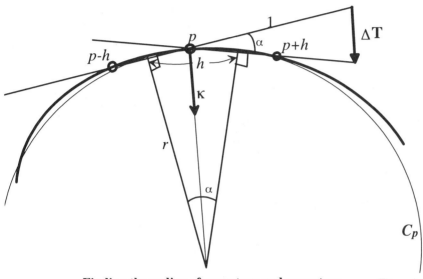

Finding the radius of curvature and curvature.   Figure A.1

Smooth Surfaces, Curvature, Geodesics, and Isometries

An (*extrinsic*) *smooth surface* is a geometric object such that, when one zooms in on its points, it is uniformly indistinguishable from a plane. When one zooms in on the point p, the plane is called the *tangent plane* (or *tangent space*) *at* p. The line which is perpendicular to the tangent plane at p is called the *normal line* (or *normal space*) *at* p. Note that at its cone point, a cone (with cone angle not equal $360°$) is not a smooth surface.

If γ is a smooth curve on the surface Σ, then the *extrinsic curvature of* γ *at* p is merely the curvature κ at p. The projection of the vector κ onto the tangent plane at p is called the *intrinsic curvature* (or *geodesic curvature*) *of* γ (written κ_g) and represents the amount that γ is intrinsically curving in Σ. The projection of κ onto the normal line at p is called the *normal curvature of* γ (written κ_n) and represents that portion of γ's curvature that is due to the curving of the surface. In vector terms, $\kappa = \kappa_g + \kappa_n$.

The curve γ is called a *geodesic in Σ* if $\kappa_g \equiv 0$ at every point of γ. In other words γ is a geodesic if it has no intrinsic curvature. This is a direct expression of "not turning" with respect to the surface. Since intrinsically the direction of the tangent line to a geodesic γ is not changing, it is natural to say that the tangent lines at two different points on γ are *parallel transports along γ*. The tangent line to γ at p is clearly a line lying in the tangent plane to Σ at p. A vector v in the tangent plane to Σ at p is said to be a *parallel transport along γ* of a vector v' in the tangent plane to Σ at p' (also on γ) if v and v' make the same angle with the tangent line to γ.

Two surfaces (smooth or not) are said to be *isometric* if they are intrinsically the same. For example, if you gently bend a piece of paper, it forms a smooth surface which is isometric to a flat sheet of paper. If you fold the paper to create a sharp crease, then it is no longer an (extrinsic) smooth surface, but it still forms a surface which is isometric to the flat surface. We may call a surface *intrinsically smooth* if it is locally isometric to an (extrinsic) smooth surface.

Theorems on Geodesics

The following are some theorems related to geodesics from the standard theory of differential geometry.

♦   *If the surface Σ is smooth and geodesically complete, then the shortest path between any two points is always a geodesic.* Here "*geodesically complete*" means that every geodesic can be extended indefinitely, that is, there are no edges or holes in the surface.

♦   *If the surface Σ is smooth (but not necessarily geodesically complete), then for any geodesic in Σ, if two points are sufficiently close, then the geodesic is the shortest path between them on the surface.*

♦   *If a (straight) ribbon is laid flat on the smooth surface Σ (that is, the ribbon is everywhere tangent to the surface along its center line) then the center line of the ribbon follows a geodesic on the surface.*

Bibliography

A. Ancient Texts

al'Khowarizmi, *Algebra*, (Robert of Chester's Latin Translation), L.C. Karpinski, ed. and trans., New York: Macmillan, 1915.
This is an English translation.

Apollonius of Perga, *Treatise on Conic Sections*, T.L. Heath, ed., New York: Dover, 1961.

Baudhayana, *Sulbasutram*, G. Thibaut, trans., S. Prakash & R. M. Sharma, ed., Bombay: Ram Swarup Sharma, 1968.

Cardano, Girolamo, *The Great Art or the Rules of Algebra*, T.R. Witmer, ed., Cambridge: MIT Press, 1968.

Euclid, *Elements*, T.L. Heath, ed., New York: Dover, 1956.

Euclid, *Optics*, H. E. Burton, trans., *Journal of the Optical Society of America*, vol. 35, no. 5, pp. 357-372, 1945.

Euclid, *Phaenomena*, in *Euclidis opera omnia*, Heinrich Menge, ed., Lipsiae: B.G. Teubneri, 1883-1916.

The Holy Bible, NIV Zondervan Bible Publishers, 1985.

Khayyam, Omar, *Algebra*, D.S. Kasir, ed., New York: Columbia Teachers College, 1931. (and New York, AMS Press, 1972.)

Khayyam, Omar, *Risâla fî sharh mâ ashkala min musâdarât Kitâb 'Uglîdis* (Arabic: Explanation of the Difficulties in Euclid's Postulates), A.I. Sabra, ed., Alexandria, Egypt: Al Maaref, 1961.
Translated in A. R. Amir-Moez, "Discussion of Difficulties" in Euclid by Omar ibn Abrahim al-Khayyami (Omar Khayyam), *Scripta Mathematica*, 24 (1958-59), pp. 275-303.

Khayyam, a paper (no title).
Translated in A. R. Amir-Moez, "A Paper of Omar Khayyam," *Scripta Mathematica*, 26(1963), pp.323-337.

Koran (Holy Qur-An), Abdullah Yusuf Ali, trans., New York: Harper Publishing, 1946.

Plato, *The Collected Dialogues*, Edith Hamilton and Huntington Carns, eds., Princeton, NJ: Bollinger, 1961.

Plotinus, *The Enneads*, Stephen McKenna, trans., Burdette, NY: Larson, 1992.

Thomas, Ivor, trans.,*Selections Illustrating the History of Greek Matheamtics*, Cambridge, MA: Harvard University Press, 1951.

B. Art and Design

Edmondson, Amy C., *A Fuller Explanation: The Synergetic Geometry of R. Buckminster Fuller*, Boston: Birkhauser, 1987.

Ernst, Bruno, *The Magic Mirror of M. C. Escher*, New York: Random House, 1976.
A revealing look at the artist and the ideas behind his work.

Ghyka, Matila, *The Geometry of Art and Life*, New York: Dover Publications, 1977.

Henderson, Linda, *The Fourth Dimension and Non-Euclidean Geometry in Modern Art*, Princeton, NJ: Princeton University Press, 1983.

Linn, Charles, *The Golden Mean: Mathematics and the Fine Arts*, Garden City, NY: Doubleday, 1974.

Miyazaki, Kojiv, *An Adventure in Multidimensional Space*, New York: John Wiley and Sons, Inc., 1983.
"The art and geometry of polygons, polyhedra, and polytopes."

Williams, Robert, *The Geometrical Foundation of Natural Structure: A Source Book of Design*, New York: Dover, 1979.

C. Differential Geometry

Dodson, C. T. J., and T. Poston, *Tensor Geometry*, London: Pitman, 1979.
   A very readable but technical text using linear (affine) algebra to study the local intrinsic geometry of spaces leading up to and including the geometry of the theory of relativity.

Dubrovin, B.A., A.T. Fomenko, S.P. Novikov, *Modern Geometry—Methods and Applications (Part I. The Geometry of Surfaces, Transformation Groups, and Fields)*, Robert G. Burns, trans., New York: Springer-Verlag, 1984.
   A well-written graduate text.

Koenderink, Jan J., *Solid Shape*, Cambridge: M.I.T. Press, 1990.
   Written for engineers and applied mathematicians, this is a discussion of the extrinsic properties of three-dimensional shapes.

Penrose, Roger, "The Geometry of the Universe," *Mathematics Today*, Lynn Steen, ed., New York: Springer-Verlag, 1978.

Spivak, Michael, *A Comprehensive Introduction to Differential Geometry*, Wilmington, DE: Publish or Perish, Inc., 1970.
   In four(!) volumes Spivak relates the subject back to the original sources.

Stahl, Saul, *The Poincaré Half-Plane*, Boston: Jones and Bartlett Publishers, 1993.

Weeks, Jeffrey, *The Shape of Space*, New York: Marcel Dekker, 1985.
   An elementary but deep discussion of the geometry on different two- and three-dimensional spaces.

D. Dimensions and Scale

Abbott, Edwin A., *Flatland*, New York: Dover Publications, Inc., 1952.
   A fantasy about 2-dimensional being in a plane encountering the third dimension.

Banchoff, Thomas and John Wermer, *Beyond the Third Dimension: Geometry, Computer Graphics, and Higher Dimensions*, New York: Springer-Verlag, 1983.

Burger, Dionys, *Sphereland*, New York: Thomas Y. Crowell Co., 1965.
A sequel to Abbott's *Flatland*.

Morrison, Phillip, and Phylis Morrison, *Powers of Ten: About the Relative Size of Things in the Universe*, New York: Scientific American Books, Inc., 1982.

Rucker, Rudy, *The Fourth Dimension*, Boston: Houghton Mifflin Co., 1984.
A history aand description of various ways that people have considered the fourth dimension.

Rucker, Rudy, *Geometry, Relativity and the Fourth Dimension*, New York: Dover, 1977.

E. Fractals

Lauwerier, Hans, *Fractals: Endlessly Repeated Geometric Figures*, Princeton, NJ: Princeton University Press, 1991.

Mandelbrot, Benoit B., *The Fractal Geometry of Nature*, New York: W.H. Freeman and Company, 1983.
The book that started the popularity of fractal geometry.

F. Geometry in Different Cultures

Albarn, K., Jenny Miall Smith, Stanford Steele, Dinah Walker, *The Language of Pattern*, New York: Harper & Row, 1974.
An enquiry inspired by Islamic decoration.

Ascher, Marcia, *Ethnomathematics: A Multicultural View of Mathematical Ideas*, Pacific Grove, CA: Brooks/Cole, 1991.

Bain, George, *Celtic Arts: The Methods of Construction*, London: Constable, 1977.

Gerdes, Paulus, *Geometrical Recreations of Africa*, Maputo, Mozambique: African Mathematical Union and Higher Pedagogical Institute's Faculty of Science, 1991.

Kline, Morris, *Mathematics in Western Culture*, New York: Oxford University Press, 1961.

Pinxten, R., Ingrid van Dooren, Frank Harvey, *The Anthropology of Space*, Philadelphia: University of Pennsylvania Press, 1983.
Concepts of geometry and space in the Navajo culture.

Zaslavsky, Claudia, *Africa Counts*, Boston: Prindle, Weber, and Schmidt, Inc., 1973.
A presentation of the mathematics in African cultures.

G. History

Beckmann, Peter, *A History of* π , Boulder, CO: The Golem Press, 1970.
A well-written enjoyable book about all aspects of π.

Bold, Benjamin, *Famous Problems of Geometry and How to Solve Them*, New York: Dover Publications, Inc., 1969.

Berggren, *Episodes in the Mathematics of Medieval Islam*, New York: Springer-Verlag, 1986.

Calinger, Ronald, *Classics of Mathematics*, Englewood Cliffs, NJ: Prentice Hall, 1995.
Mostly a collection of original sources in Western mathematics.

Carroll, Lewis, *Euclid and His Modern Rivals*, New York: Dover Publications, Inc., 1973.
Yes! Lewis Carroll of Alice in Wonderland fame was a geometer. This book is written as a drama; Carroll has Euclid defending himself against modern critics.

Eves, Howard, *Great Moments in Mathematics (after 1650)*, Dolciani Mathematical Expositions, Vol. 7, Washington, DC: M.A.A., 1981.

Joseph, George, *The Crest of the Peacock*, New York: I.B. Tauris, 1991.
A non-Eurocentric view of the history of mathematics.

Kline, Morris, *Mathematical Thought from Ancient to Modern Times*, Oxford: Oxford University Press, 1972.
A complete Eurocentric history of mathematical ideas.

Newell, Virginia K. (ed.) *Black Mathematicians and Their Works*, Ardmore, PA: Dorrance, 1980.

Richards, Joan, *Mathematical Visions*, Boston: Academic Press, 1988. "The pusuit of geometry in Victorian England."

Seidenberg, A., The Ritual Origin of Geometry, *Archive for the History of the Exact Sciences*, 1(1961), pp. 488-527.

Smeltzer, Donald, *Man and Number*, New York: Emerson Books, 1958. History and cultural aspects of mathematics.

Valens, Evans G., *The Number of Things: Pythagoras, Geometry and Humming Strings*, New York: E.P. Dutton and Company, 1964. This is a book about ideas and is not a textbook. Valens leads the reader thru dissections, golden mean, relations between geometry and music, conic sections, etc.

H. Linear Algebra and Geometry

Banchoff, T., and J. Wermer, *Linear Algebra through Geometry*, New York: Springer-Verlag, 1983. For several years this was the text for Cornell's undergraduate linear algebra course.

Dodson, C. T. J., and T. Poston, *Tensor Geometry,* London: Pitman, 1979. A very readable but technical text using linear (affine) algebra to study the local intrinsic geometry of spaces leading up to and including the geometry of the theory of relativity.

Murtha, James A., and Earl R. Willard, *Linear Algebra and Geometry*, New York: Holt, Reinhart and Winston, Inc., 1966. Includes affine and projective geometry.

Taylor, Walter F., *The Geometry of Computer Graphics*, Pacific Grove, CA: Wadsworth and Brooks, 1992.

I. Models, Polyhedra

Barr, Stephen, *Experiments in Topology*, New York: Crowell, 1964.

Barnette, David, *Map Colouring, Polyhedra, and the Four-Colour Problem*, Dociani Mathematical Expositions Vol. 8, Washington, DC: M.A.A., 1983.

Cundy, M.H., and A.P. Rollett, *Mathematical Models*, Oxford: Clarendon, 1961.
    Directions on how to make and understand various geometric models.

Lyusternik, L.A., *Convex Figures and Polyhedra*, Boston: Heath, 1966.

Row, T. Sundra, *Geometric Exercises in Paper Folding*, New York: Dover, 1966.
    How to produce various geometric constructions merely by folding a sheet of paper.

Senechal, Marjorie, and George Fleck, *Shaping Space: A Polyheral Approach*, Design Science Collection, Boston: Birkhauser, 1988.

J. Nature

Cook, T.A., *The Curves of Life: Being an Account of Spiral Formations and their Applications to Growth in Nature, to Science, and to Art*, New York: Dover Publications, 1979.

Ghyka, Matila, *The Geometry of Art and Life*, New York: Dover Publications, 1977.

Mandelbrot, Benoit B., *The Factal Geometry of Nature*, New York: W.H. Freeman and Company, 1983.
    The book that started the popularity of fractal geometry.

McMahon, Thomas and James Bonner, *On Size and Life*, New York: Scientific American Library, 1983.
    A geometric discussion of the shapes and sizes of living things.

Thom, Rene, *Structural Stabililty and Morphogenesis*, Redwood City, CA: Addison-Wesley, 1989.
A geometric and analytic treatment of "Catastrophe Theory."

Thompson, D'Arcy, *On Growth and Form*, Cambridge: Cambridge University Press, 1961.
A classic on the geometry of the natural world.

K. Non-Euclidean Geometries (Mostly Hyperbolic)

Greenberg, Marvin J., *Euclidean and Non-Euclidean Geometries: Development and History*, New York: Freeman, 1980
This is a very readable textbook that includes some philosophical discussions.

Petit, Jean-Pierre, *Euclid Rules OK? The Adventures of Archibald Higgins*, London: John Murray, 1982.
A pictoral, visual tour of non-Euclidean geometries.

Millman, Richard S., and George D. Parker, *Geometry: A Metric Approach with Models*, New York: Springer-Verlag, 1981.
A modern formal axiomatic approach.

Nikulin, V.V., and I.R. Shafarevich, *Geometries and Groups*, Berlin: Springer-Verlag, 1987.
Using transformation groups to study spherical, hyperbolic, and toroidal geometries.

Schwerdtfeger, Hans, *Geometry of Complex Numbers: Circle Geometry, Moebius Transformation, Non-Euclidean Geometry*, New York: Dover Publications, Inc., 1979.

Ryan, Patrick J., *Euclidean and Non-Euclidean Geometry: An Analytic Approach*, Cambridge: Cambridge University Press, 1986.

L. Philosophy

Benacerraf, Paul, and Hilary Putman, *Philosophy of Mathematics: Selected Readings*, Cambridge: Cambridge University Press, 1964.

Hofstadter, Douglas R., *Gödel, Escher, Bach: An Eternal Golden Braid*, New York: Basic Books, 1979.

Lachterman, David Rapport, *The Ethics of Geometry: A Genealogy of Modernity*, New York: Routledge, 1989.

Lakatos, I., *Proofs and Refutations*, Cambridge: Cambridge University Press, 1976.

Stein, Charles (ed.), *Being = Space X Action*, IO Vol. 41, Berkeley, CA: North Atlantic Books, 1988.
"Searches for Freedom of Mind through Mathematics, Art, and Mysticism."

Tymoczko, Thomas, *New Directions in the Philosophy of Mathematics*, Boston: Birkhauser, 1986.

M. Spherical and Projective Geometry

Todhunter, Isaac, *Spherical Trigonometry*, London: Macmillan, 1886.
All you want to know, and more, about trigonometry on the sphere. Well written with nice discussions of surveying.

Whicher, Olive, *Projective Geometry: Creative Polarities in Space and Time*, London: Rudolf Steiner Press, 1971.
Projective geometry is the geometry of perception and prospective drawings.

N. Symmetry and Groups

Budden, F.J., *Fascination of Groups*, Cambridge: Cambridge University Press, 1972.
This is a fascinating book that relates algebra (groups) to geometry, music, and so forth, and has a nice description of symmetry and patterns.

Bunch, Bryan H., *Reality's Mirror: Exploring the Mathematics of Symmetry*, New York: John Wiley, 1989.

Burn, R.P., *Groups: A Pathway to Geometry*, Cambridge: Cambridge University Press, 1985.

Weyl, Hermann, *Symmetry*, Princeton, NJ: Princeton University Press, 1952.
A readable discussion of all mathematical aspects of symmetry especially its relation to art and nature — nice pictures. Weyl is a leading mathematician of this century.

O. Surveys and General Expositions

Davis, P.J., and R. Hersh, *The Mathematical Experience*, Boston: Birkhauser, 1981.
A very readable collection of essays by two present-day mathematicians. I think every mathematics major should own this book.

Ekeland, Ivar, *Mathematics and the Unexpected*, Chicago: University of Chicago Press, 1988.

Gaffney, Matthew P. and Lynn Arthur Steen, *Annotated Bibliography of Expository Writing in the Mathematical Sciences*, Washington, DC: M.A.A., 1976.

Gamow, George, *One, Two, Three ... Infinity*, New York: Bantam Books, 1961.
A well-written journey through mathematical ideas.

Guillen, Michail, *Bridges to Infinity: The Human Side of Mathematics*, Los Angeles: Jeremy P. Tarcher, 1983.

Hilbert, David, and S. Cohn-Vossen, *Geometry and the Imagination*, New York: Chelsea Publishing Co., 1983.
They state "it is our purpose to give a presentation of geometry, as it stands today [1932], in its visual, intuitive aspects." It includes an introduction to differential geometry, symmetry, and patterns (they call it "crystallographic groups"), and the geometry of spheres and other surfaces. Hilbert is the most famous mathematician of the first part of this century.

Honsberger, Ross, *Mathematical Gems*, Dolciani Mathematical Expositions, Vol. 2, Washington, DC: M.A.A., 1973.

Honsberger, Ross, *Mathematical Gems II*, Dolciani Mathematical Expositions, Vol. 4, Washington, DC: M.A.A., 1976.

Honsberger, Ross, *Mathematical Morsels*, Dolciani Mathematical Expositions, Vol. 1, Washington, DC: M.A.A., 1978.

Honsberger, Ross, *Mathematical Plums*, Dolciani Mathematical Expositions, Vol. 4, Washington, DC: M.A.A., 1979.
Expository stories about mathematics.

Jester, Norton, *The Dot and the Line: A Romance in Lower Mathematics*, New York: Random House, 1963.
A mathematical fable.

Lang, Serge, *The Beauty of Doing Mathematics: Three Public Dialogues*, New York: Springer-Verlag, 1985.
Expository work by a famous mathematician.

Lieber, Lillian R., *The Education of T.C. Mits (The Celebrated Man in the Street)*, New York: W.W. Norton, 1972.
A mathematical fantasy.

Péter, Rozsa, *Playing with Infinity*, New York: Dover Pubslishing, Inc., 1961.
"Mathematical explorations and excursions."

Steen, Lynn Arthur (ed.), *Mathematics Tomorrow*, New York: Springer-Verlag, 1981.
Expository essays.

Steen, Lynn Arthur (ed.), *Mathematics Today: Twelve Informal Essays*, New York: Springer-Verlag, 1978.

Stewart, Ian, *The Problems of Mathematics*, Oxford: Oxford University Press, 1987.

P. Texts

Coxeter, H.S.M., *Introduction to Geometry*, New York: Wiley, 1969.
This is a collection of diverse topics including non-Euclidean geometry, symmetry, patterns, and much, much more. Coxeter is one of the foremost living geometers.

Eves, Howard, *A Survey of Geometry,* Vol. 1, Boston: Allyn & Bacon, 1963.

Eves, Howard, *Modern Elementary Geometry*, Boston: Jones and Bartlett Publishing, 1992.

Jacobs, Harold R., *Geometry*, San Fransisco: W.H. Freeman and Co., 1974.
A high-school-level text based on guided discovery.

Serra, Michael, *Discovering Geometry: An Inductive Approach*, Berkeley, CA: Key Curriculum Press, 1989.

Q. Miscellaneous

Ho, Chung-Wu, "Decomposition of a Polygon into Triangles," *Mathematical Gazette*, 60 (1976), pp.132-134.

Kempe, A.B., *How to Draw a Straight Line*, London: Macmillan, 1877.

Sah, C.H., *Hilbert's Third Problem: Scissors Congruence*, London: Pitman, 1979.

Index

A

AAA = Angle-Angle-Angle, 83, 84,
    153, 154
al'Khowarizmi, 126, 127, 130, 177
algebra, 112, 122, 126-128, 134, 139,
    140
    abstract groups, 157, 160, 164
angle, 8, 22, 24, 140, 141, 144, 149,
    153, 154
    bisector, 68, 69
    cone, 30, 31, 35, 37, 40, 175
    definitions, 23, 24, 26
    dihedral, 125, 140, 166-170
    directed, 24
    right, 28, 36, 85, 88, 90, 110, 151
    solid, 166, 167, 168, 170, 171
    sum of, 54, 57, 62, 79, 82, 85, 86,
        93, 94, 104
    vertical, 22, 24
antipodal, 69, 152
Apollonius, 132, 177
area, 24, 57, 104, 107, 112, 115, 117,
    127, 141, 143, 167, 168, 171
    and dissection theory, 115, 121
ASA = Angle-Side-Angle, 47, 52, 53,
    54, 67, 68, 70, 152, 153, 154
ASS = Angle-Side-Side, 72, 73, 74,
    81
axioms = postulates, 36, 48, 91, 139
axis, 6, 8

B

Baudhayana, 112, 118, 121, 129, 177

biangle = lune, 57, 107, 111
Bible, 1, 177
Bolyai, 91

C

Cardano, 137, 138, 139, 177
chart, 141
circle, 6, 7, 10, 11, 14, 15, 17, 19, 20,
    24, 67, 68, 69, 72, 88, 97, 115,
    116, 131, 132, 135, 146, 147,
    152, 155, 174
    great, 34, 69
    intrinsic center, 68, 146, 152, 153
    intrinsic radius, 68, 146, 147
    osculating, 174
Clifford parallels, 102
complete, 176
Completeness Axiom, 113, 115
complex numbers, 138, 139
cone, 10, 20, 28-37, 70, 80, 140, 175
conformal mappings, 144
congruent, 23, 24, 47, 49, 52, 67, 70,
    72, 74, 81, 83, 110, 125, 152,
    157, 160, 162, 163, 165, 166,
    167, 168, 170
conic sections, 126, 128, 131, 132,
    134, 135, 138
cosets, 165
covariant (= intrinsic) differentiation,
    59
covering space, 52, 56, 70
    branch points, 39, 43, 44
    of cone, 39, 40, 43
    of cylinder, 37-40